Sub-threshold Design for Ultra Low-Power Systems

SERIES ON INTEGRATED CIRCUITS AND SYSTEMS

Anantha Chandrakasan, Editor
Massachusetts Institute of Technology
Cambridge, Massachusetts, USA

Published books in the series:

A Practical Guide for SystemVerilog Assertions
Srikanth Vijayaraghavan and Meyyappan Ramanathan
2005, ISBN 0-387-26049-8

Statistical Analysis and Optimization for VLSI: Timing and Power
Ashish Srivastava, Dennis Sylvester and David Blaauw
2005, ISBN 0-387-25738-1

Leakage in Nanometer CMOS Technologies
Siva G. Narendra and Anantha Chandrakasan
2005, ISBN 0-387-25737-3

Thermal and Power Management of Integrated Circuits
Arman Vassighi and Manoj Sachdev
2005, ISBN 0-398-25762-4

High Performance Energy Efficient Microprocessor Design
Vojin Oklobdzija and Ram Krishnamurthy (Eds.)
2006, ISBN 0-387-28594-6

Abstraction Refinement for Large Scale Model Checking
Chao Wang, Gary D. Hachtel and Fabio Somenzi
2006, ISBN 0-387-28594-6

Alice Wang
Benton H. Calhoun
Anantha P. Chandrakasan

Sub-threshold Design for Ultra Low-Power Systems

Alice Wang
Texas Instruments
12500 TI Boulevard, MS 8723
Dallas, TX 75243
U.S.A.

Benton H. Calhoun
Massachusetts Institute of Technology
50 Vassar Street, 38-107
Cambridge, MA 02139
U.S.A.

Anantha P. Chandrakasan
Massachusetts Institute of Technology
Department of Electrical Engineering
 & Computer Science
50 Vassar Street, 38-107
Cambridge, MA 02139
U.S.A

Sub-threshold Voltage Circuit Design for Ultra Low Power Systems

Library of Congress Control Number: 2006925629

ISBN-10: 0-387-33515-3   ISBN-10: 0-387-34501-9 (e-book)
ISBN-13: 9780387335155   ISBN-13: 9780387345017 (e-book)

Printed on acid-free paper.

© 2006 Springer Science+Business Media, LLC
All rights reserved. This work may not be translated or copied in whole or in part without the written permission of the publisher (Springer Science+Business Media, LLC, 233 Spring Street, New York, NY 10013, USA), except for brief excerpts in connection with reviews or scholarly analysis. Use in connection with any form of information storage and retrieval, electronic adaptation, computer software, or by similar or dissimilar methodology now known or hereafter developed is forbidden.
The use in this publication of trade names, trademarks, service marks and similar terms, even if they are not identified as such, is not to be taken as an expression of opinion as to whether or not they are subject to proprietary rights.

Printed in the United States of America.

9 8 7 6 5 4 3 2 1

springer.com

Preface

Although energy dissipation has improved with each new technology node, because SoCs are integrating tens of million devices on-chip, the energy expended per operation has become a critical consideration in digital and analog integrated circuits. The focus of this book is sub-threshold circuit design, which involves scaling voltages below the device thresholds. In this region, the energy per operation can be reduced by an order of magnitude compared to conventional operation but at the cost of circuit performance. In many emerging applications such as self-powered RFID, wireless sensors networks, and portable devices (PDAs, medical monitoring, etc.), the overall battery lifetime is the primary design metric. Sub-threshold design can also be applied to burst mode applications (e.g., a cell-phone processor) where the process spends a significant amount of time in the standby mode. The supply voltage can be reduced to the deep sub-threshold region, dramatically saving power in logic and memory.

Extremely low-power design was first explored in the 1970s for the design of applications such as wristwatch and calculator circuits. Dr. Eric Vittoz pioneered the design and modeling of weak-inversion circuits. In this book, Eric provides his perspective on the evolution of sub-threshold circuit design. Dr. Eric Vittoz and Dr. Christian Enz introduce key models necessary for the design and optimization of weak inversion circuits. Design using weak inversion has been widely adopted in analog circuits, and Eric introduces the key design considerations. His contributions and perspectives are critical to the completeness of this book. We are grateful for his insights.

Ultra-low-voltage CMOS digital operation was demonstrated by Prof. James Meindl and Dr. Richard Swanson in a Journal of Solid-State Circuits paper (April 1972); they predicted CMOS logic operating at a supply voltage of $8kt/q \approx 200\text{mV}$ at room temperature and derived the fundamental limits of voltage scaling [1]. This is a key result for low-voltage logic digital design and is an inspiration for many of the results described in this book.

This book focuses on the design of ultra-low-voltage digital circuits ($<$ 0.4V) in scaled technologies. Sub-threshold logic in scaled technology also has potential usefulness in high-performance applications. This book introduces the key challenges associated with sub-threshold design including circuit modeling, digital logic design (sizing, logic style selection, optimum supply voltage operation, etc.), memory design, and analog design. In scaled technologies and at low-voltages, the energy contribution due to sub-threshold leakage cannot be ignored. An optimum supply voltage for digital operation is derived that balances leakage and switching energy.

Conventional design approaches for logic scale well into the sub-threshold region, however, proper device sizing is necessary (especially in sequential blocks). The key challenge in the design of ultra-low-voltage digital systems is memory operation. Unfortunately, conventional memory circuits do not easily scale, so new cell-design along with architectures and I/O circuits are required. The energy must be minimized while keeping the cell area overhead to a minimum. This book highlights the importance of design methods that account for process variations, which have a larger impact in low-voltage operation.

The results described in this text book were primarily developed as a part of Ph.D. thesis research at the Massachusetts Institute of Technology [2][3]. The book provides a first step towards the design of ultra-low-voltage circuits. We expect continued development in this area and widespread use of sub-threshold design in many emerging applications.

Alice Wang *Texas Instruments, Inc.*
Benton H. Calhoun *University of Virginia*
Anantha P. Chandrakasan *Massachusetts Institute of Technology*

List of Contributors

Eric A. Vittoz, Swiss Institute of Technology Lausanne (EPFL)

Eric A. Vittoz received his E.E. degree from Polytechnical School University of Lausanne in 1961 and his Ph.D. from EPFL (Swiss Institute of Technology Lausanne) in 1969. He joined the Watchmakers Electronic Center (CEH) in 1962. He was head of the Advanced Circuit Department in 1967, then Vice Director and head of the Applications Division in 1971. In 1984, he took the responsibility of the Circuits and Systems Research Division of CSEM (Swiss center for Electronics and Microtechnology), and was Executive VP in 1991, head of Integrated Circuits and Systems, then head of Advanced Microelectronics in 1997, before retiring in 2004. Since 1975, he has been teaching analog circuit design and supervising students at EPFL, where he became professor in 1982.

Dr. Vittoz has been involved in the formation of the IEEE Solid-State Circuits Society and was a member of its AdCom from 1996 to 1999. He has authored or co-authored more than 140 papers and holds 26 patents in the fields of very low-power microelectronics, compact transistor modeling, analog CMOS circuit design and biology-inspired analog VLSI. A Life Fellow of the IEEE, he is the recipient of the 2004 IEEE Solid-State Circuits Award.

Joyce Kwong, Massachusetts Institute of Technology

Joyce Kwong received the BASc degree in computer engineering from University of Waterloo in 2004. She is currently pursuing the MS and PhD degrees in electrical engineering at Massachusetts Institute of Technology. Her research interests include sub-threshold circuits, statistical modeling and design techniques for process variation.

Christian C. Enz, Swiss Center for Electronics and Microtechnology (CSEM)

Christian Enz, PhD, Swiss Federal Institute of Technology (EPFL), 1989, is currently VP at the Swiss Center for Electronics and Microtechnology (CSEM) in Neuchtel, Switzerland, where he is head of the Microelectronics Division. Since 1999, he is also a Professor at EPFL, where he lectures and supervises students in the field of analog and RF IC design. Prior to joining the CSEM, he was a Principal Senior Engineer at Conexant (formerly Rockwell Semiconductor Systems), Newport Beach, CA. His technical interests and expertise are in the field of wireless sensor networks, very low-power analog and RF IC design and semiconductor device modeling. He is one of the developers of the EKV MOS transistor model and author of the book "Charge-Based MOS Transistor Modeling - The EKV Model for Low-Power and RF IC Design" (Wiley, 2006). He is the author and co-author of more than 130 scientific papers and has contributed to numerous presentations and advanced engineering courses.

Contents

1	**Introduction**		1
	1.1 Energy-Constrained Applications		2
		1.1.1 Micro-sensor Networks and Nodes	2
		1.1.2 Radio Frequency Identification (RFID)	3
		1.1.3 Low-power Digital Signal Processor (DSP) and Microcontroller Units (MCU)	3
	1.2 System requirements		4
		1.2.1 Battery Lifetimes	4
		1.2.2 Energy Harvesting	4
	1.3 Book Summary		5
2	**Origins of Weak Inversion (or Sub-threshold) Circuit Design**		7
	by Eric A. Vittoz		
3	**Survey of Low-voltage Implementations**		11
	3.1 Technology Scaling		11
	3.2 Low-voltage Logic Designs		14
	3.3 History of Minimum Voltage		17
	3.4 History of Minimum Energy		20
	3.5 Survey of Sub-threshold CMOS Circuits		22
4	**Minimizing Energy Consumption**		25
	4.1 Energy-Performance Contours		25
		4.1.1 Variable Activity Factor Circuit	25
		4.1.2 Energy-Performance Contours	26
		4.1.3 Activity Factor	28
	4.2 Modeling Minimum Energy Consumption		30
		4.2.1 Sub-Threshold Leakage Current Models	30
		4.2.2 Other Components of Current	34
		4.2.3 Minimum Energy Point Model	35

	4.3	Minimum Energy Point Dependencies 42
		4.3.1 Operating Scenario 42
		4.3.2 Temperature 45
		4.3.3 Architecture 46

5 EKV Model of the MOS Transistor 49
by Eric A. Vittoz and Christian C. Enz

- 5.1 Introduction and Definitions 49
- 5.2 Density of Mobile Charge 50
  - 5.2.1 Threshold Function 50
  - 5.2.2 Approximation in Strong Inversion 52
  - 5.2.3 General Case 53
  - 5.2.4 Approximation in Weak Inversion 54
- 5.3 Drain Current and Modes of Operation 54
  - 5.3.1 Charge-Current Relationship 54
  - 5.3.2 Forward and Reverse Components 55
  - 5.3.3 General Current Expression 56
  - 5.3.4 Modes of Operation and Inversion Coefficient 57
  - 5.3.5 Output Characteristics and Saturation Voltage 58
  - 5.3.6 Weak Inversion Approximation 59
- 5.4 Small-Signal Model 59
  - 5.4.1 Transconductances 59
  - 5.4.2 Residual Conductance in Saturation and Maximum Voltage Gain 61
  - 5.4.3 Small-Signal AC Model 61
- 5.5 Transistor Operated As a Pseudo-Resistor 62
- 5.6 Noise 63
  - 5.6.1 Noise model 63
  - 5.6.2 Channel Noise 64
  - 5.6.3 Interface Noise 65
  - 5.6.4 Total Noise 65
- 5.7 Temperature effects 65
- 5.8 Non-ideal effects 67
  - 5.8.1 Mismatch 67
  - 5.8.2 Polysilicon Gate Depletion 69
  - 5.8.3 Band Gap Widening 70
  - 5.8.4 Gate Leakage 71
  - 5.8.5 Drain-Induced Barrier Lowering (DIBL) 72

6 Digital Logic 75
- 6.1 Inverter Operation in Sub-threshold 75
  - 6.1.1 Sub-threshold Inverter Delay 75
  - 6.1.2 Sub-threshold Voltage Transfer Characteristics (VTCs). 77
  - 6.1.3 Inverter Sizing for Minimum Energy 82
- 6.2 Sub-threshold CMOS Standard Cell Library 83

		6.2.1	Parallel Devices 84
		6.2.2	Stacked Devices 86
		6.2.3	Flip-flops... 87
		6.2.4	Ratioed Circuits 88
		6.2.5	Measured Results from Test Chip 89
	6.3	Logic Families in Sub-threshold 92	
		by Joyce Kwong	
		6.3.1	Process Variation in Sub-threshold Logic 92
		6.3.2	Evaluating Logic Styles in the Context of Variations ... 95

7 Sub-threshold Memories 103
7.1 Register Files 103
7.1.1 Write Port and Memory Cell........................ 104
7.1.2 Read Bitline Architectures.......................... 105
7.1.3 Sub-threshold Register File 114
7.2 Sub-threshold SRAM 115
7.2.1 SRAM Overview 116
7.2.2 6-Transistor SRAM Bitcell in Sub-threshold 123
7.2.3 Write Operation 124
7.2.4 Read Operation 127
7.2.5 Static Noise Margin in sub-threshold................. 129
7.2.6 A Sub-threshold Bit-cell Design 131
7.2.7 65nm Sub-threshold SRAM Test Chip 139

8 Analog Circuits in Weak Inversion 147
by Eric A. Vittoz
8.1 Introduction 147
8.2 Minimum Saturation Voltage 148
8.2.1 Current Mirrors 148
8.2.2 Cascode Mirrors................................... 148
8.2.3 Low-Voltage Amplifiers 150
8.3 Maximum Transconductance-to-Current Ratio 151
8.3.1 Differential Pair 151
8.3.2 Single-Stage Operational Transconductance Amplifiers (OTA) 154
8.4 Exponential Characteristics 156
8.4.1 Voltage and Current Reference 156
8.4.2 Amplitude Regulator 157
8.4.3 Translinear Circuits 158
8.4.4 Log-Domain Filters [177, 178, 179, 180] 161
8.5 Pseudo-Resistor 162
8.5.1 Analysis of Circuits................................ 162
8.5.2 Emulation of Variable Resistive Networks 163

XII Contents

9 System Examples 167
    9.1 A Sub-threshold FFT Processor 167
        9.1.1 The Fast Fourier Transform 168
        9.1.2 Energy-Aware Architectures 169
        9.1.3 Minimum Energy Point Analysis 170
        9.1.4 Measurements 171
    9.2 Ultra-Dynamic Voltage Scaling 173
        9.2.1 DVS and Local Voltage Dithering 174
        9.2.2 Ultra-Dynamic Voltage Scaling (UDVS) Test Chip..... 178
        9.2.3 UDVS System Considerations 184

A Acronyms 191

References 193

Index 207

1
Introduction

This book focuses on sub-threshold, or weak inversion, circuit design. Digital circuits operating in sub-threshold use a supply voltage that is less than the threshold voltages of the transistors. In this region of operation, they consume less energy for active operation and dissipate less leakage power than higher voltage alternatives, but they operate more slowly. Until fairly recently, the emphasis on maximizing operational frequency in digital circuits dominated to the point that sub-threshold operation received very little attention.

A recent explosion in applications that benefit from low energy operation has carved out a significant niche for sub-threshold circuits. Gadgets that for years were tied down or wired up have severed the cords and broken loose. This trend toward portability has created two classes of applications for which sub-threshold circuits are well-suited. The first is severely energy constrained systems. For systems in this class, conserving energy is the primary constraint, and speed of operation is largely irrelevant. Instead, minimizing energy per operation becomes the primary goal. Sub-threshold circuits are ideal for these types of applications. The second class of applications consists of portable devices that require high performance for part of the time but that also spend significant fractions of their operation doing non-performance critical tasks. The mobile phone is a perfect example since it often enters periods of near-idle computation to wait for input from the user or the wireless link. For these type of applications, Ultra-Dynamic Voltage Scaling (UDVS) is a good solution. UDVS is a technique which allows the same circuit to operate at high-voltage/high-frequency for performance critical applications and at sub-threshold/low-frequency for energy-constrained applications.

Mobile technologies are enabled by wireless communication and by portable power supplies. The latter most often takes the form of a battery. Improvements in both of these categories have led to remarkable changes in portable electronics. For example, the first commercial mobile phone weighed 16 ounces and had a half-hour of talk time. Now more than 10 years later, the cell phone weighs less than 3 ounces and provides up to 4-6 hours of talk time. Many cell phones also add features including email, MP3 playback, gaming, digital TV and movies. The increase of features in portable electronics along with the decreasing form factor places a heavy burden on the battery to supply more energy from less volume. However, the energy density of batteries has only doubled every five to 20 years, depending on the battery chemistry. At the same time, prolonged refinement of any chemistry yields diminishing returns. As a result, the energy burden shifts to the circuits, which must reduce the energy consumed during each operation in order to extend the system lifetime.

This chapter briefly explores examples of energy-constrained applications and the requirements for these systems. Although best suited to these applications, sub-threshold operation can also apply to other portable devices during periods of non-performance critical operation. The remainder of the book looks in detail at sub-threshold operation as a potential approach to these applications.

1.1 Energy-Constrained Applications

There is an emerging set of applications for which energy consumption is the key metric. These severely energy-constrained applications generally have low activity rates and low speed requirements, but the system is required to have long battery lifetimes (e.g. > 1 year). Ideally, the power consumption of these systems will decrease to the point that they can harvest energy from their environments and have theoretically unlimited lifetimes.

1.1.1 Micro-sensor Networks and Nodes

A micro-sensor node refers to physical hardware that provides sensing, computation, and communication functionality. A wireless micro-sensor network consists of tens to thousands of distributed nodes that sense and process data and relay the results to the end-user. Proposed applications for micro-sensor networks include habitat monitoring [4][5][6], health [7], structural monitoring [8][9][10], and automotive sensing [7].

The performance requirements for microsensor nodes in these applications are very low. The rate at which data changes for environmental or health monitoring, for example, is on the order of seconds to minutes, so the performance achieved even in sub-threshold is more than adequate. Most micro-sensor nodes duty cycle, or shutdown unused components whenever possible. Although duty cycling helps to extend sensor network lifetimes, it does not remove the energy constraint placed by the battery. Most microsensor node applications require very long battery lifetimes because it is not possible to recharge or replace batteries frequently. Thus, microsensor networks are a compelling platform that showcases the need for new low-energy design techniques.

1.1.2 Radio Frequency Identification (RFID)

Radio Frequency Identification (RFID) is another application which requires extremely low energy consumption [11]. RFID is used to automatically identify objects through RFID tags which are attached to the object. The RFID tag is able to transmit and receive information wirelessly using radio-frequencies. An RFID tag contains a limited amount of digital processing logic along with an antenna and communication circuits.

Although the concept of RFID has been in existence for many years, recently it has become more popular as a means to efficiently control large scale supply chains. Its success has spawned many more applications for RFID tags including medical implants, pet identification, smart credit cards and smart keys for automobiles.

There are two main types of RFID tags. An active RFID tag communicates with the reader by transmitting data. Active tags frequently require batteries to supply the energy for transmission, and the extra energy from the battery also allows for extended processing and longer range of communication. A passive RFID tag communicates with the reader by modulating the load that the reader sees. This indirect means of communication requires less energy, so passive tags often operate on energy that is converted from the received signal. Passive nodes are usually smaller as a result, and their lifetimes are not limited by energy.

Reducing the digital processing power would benefit both types of tags. For passive tags, the power is constrained by the ability to utilize the converted energy from the antenna. If the digital logic power dissipation can be reduced, then the distance from the reader to the tag can increase since less transmitted power has to reach the tag. For active tags, minimizing the digital logic power leads to both increased transmission range and/or longer battery lifetimes.

1.1.3 Low-power Digital Signal Processor (DSP) and Microcontroller Units (MCU)

Most portable electronics that are used for consumer applications require a low-power DSP or MCU. Even if the amount of processing is limited, it must be done efficiently. In a variety of applications, the Texas Instruments (TI) C5xx family of DSPs or the TI MSP430 family of MCUs have been used successfully for portable measurement, metering and instrumentation.

Also, there are portable applications that take advantage of a wide range of performance needs to reduce energy consumption and extend the system lifetime. For example, a mobile phone may be on standby for most of its lifetime but then require increased processor MIPS (Million Instructions per Second) when the user makes a call or runs an application (e.g. games, digital TV). In these applications, the DSP and MCU need a wide dynamic range of power/performance. They should both minimize energy when in standby or low activity modes and maximize performance when in high activity modes. Design considerations for both spaces are needed to optimize these devices.

1.2 System requirements

In this section, we motivate the need for sub-threshold circuit design based on the system requirements for severely energy-constrained systems.

1.2.1 Battery Lifetimes

For some micro-sensor applications, a limited lifetime is sufficient, and a non-rechargeable battery is the logical choice. The small size of the sensor nodes imply limited physical space for batteries. However, a 1cm^3 Lithium battery can continuously supply 10μW of power for five years [12].

Since the standby power alone of most large digital systems exceeds this number, energy conservation strategies are essential for achieving the lifetimes necessary for viable applications [13]. Sub-threshold circuit design, which dissipates much less leakage power and consumes less active energy than strong inversion circuits, provides useful computation at the power levels required for extended battery lifetimes.

1.2.2 Energy Harvesting

Many of the applications mentioned in Section 1.1 would benefit from unbounded lifetimes in an environment where changing batteries is impractical or impossible. These types of applications require a renewable energy source. The concept of energy harvesting or energy scavenging involves converting ambient energy from the environment into electrical energy to power circuits or to recharge a battery.

Table 1.1. Examples of power densities for potential energy harvesting mechanisms [14] (© 2005 IEEE)

Technology	Power Density (μW/cm^2)
Vibration - electromagnetic [15]	4.0
Vibration - piezoelectric [12]	500
Vibration - electrostatic [16]	3.8
Thermoelectric (5°C difference) [17]	60
Solar - direct sunlight [18]	3700
Solar - indoor [18]	3.2

The most familiar sources of ambient energy include solar power, thermal gradients, radio-frequency (RF), and mechanical vibration. Table 1.1 gives a comparison of some energy harvesting technologies [14]. Power per area is reported because the thickness of these devices is typically dominated by the other two dimensions. The power available from these sources is highly dependent on the nodes' environment at any given time. However, these examples show that it is reasonable to expect 10's of microwatts of power to be harvested from ambient energy. Thus, researchers agree that micro-sensor nodes must keep average power consumption in the 10-100μW range to enable energy scavenging [19][20]. Coupling energy-harvesting techniques with some form of energy storage can theoretically extend system lifetimes indefinitely. Clearly, this type of system will be much more effective when coupled with the significant power and energy savings made possible by sub-threshold operation.

1.3 Book Summary

In this chapter we showed the motivation for sub-threshold design in future energy-constrained applications. The battery capacity of portable systems is not keeping up with the SoC requirements. Also, new wireless applications are emerging that require near infinite lifetimes. One way to extend battery lifetime is to run at microwatt power levels. If average power consumption is sufficiently low, then energy harvesting becomes possible. Sub-threshold circuit design provides a solution where a circuit can operate at the voltage supply that minimizes energy dissipation.

This book covers various aspects of sub-threshold design.

Chapter 2 is a contributed chapter by E. Vittoz, a pioneer in sub-threshold design. He gives his unique perspective on the origins of sub-threshold circuit design in which he had an important role. Then, in Chapter 3, we survey low-voltage, low-power circuit designs from as early as the 1960's. We show the trends in voltage scaling for processors, DSP's and research test chips and identify results that pushed the minimum voltage limit. Chapter 4 explores the concept of optimal energy dissipation. Minimizing energy dissipation is the primary design objective for many energy-constrained systems. We explore operating at the optimum point for minimum energy dissipation over various system parameters such as supply voltage, threshold voltage, activity factor, workload, duty cycle and temperature. A thorough discussion of sub-threshold leakage current modeling is presented in Chapter 5, which is contributed and written by E. Vittoz and C. Enz. The Enz, Krummenacher, and Vittoz (EKV) model that is presented is intended for usage in low-voltage and low-current digital and analog design. Chapter 6 discusses digital logic operation in the sub-threshold region. First the inverter in sub-threshold is analyzed, and then complex CMOS logic circuits are analyzed. In addition, different logic families are considered in a section contributed by J. Kwong. In Chapter 7, sub-threshold memories are discussed. We explore traditional SRAM designs that operate in strong inversion but fail in sub-threshold. Understanding the limitations of SRAM write and read helps us to design new bitcells and read/write circuits for sub-threshold operation. Chapter 8, which is contributed by E. Vittoz, moves on to sub-threshold analog circuit design and highlights common circuits and design techniques that take advantage of specific characteristics of weak inversion operation. The last chapter presents two system examples that incorporate ideas presented in the preceding chapters. The first system example is a Fast Fourier Transform (FFT) processor that operates down to 180mV and demonstrates optimal energy dissipation. The second system demonstrates UDVS, which allows a system to dynamically scale from strong inversion down to sub-threshold. The contributed chapters (Chapters 2, 5 and 8) use different terminology than the remainder of the book. A large portion of the content of this book comes from the Ph.D. theses of Dr. Wang and Dr. Calhoun, which were written at the Massachusetts Institute of Technology.

2
Origins of Weak Inversion (or Sub-threshold) Circuit Design

by Eric A. Vittoz

The state of weak inversion at the silicon surface in a metal-insulator-silicon structure was already implicitly mentioned as the "parabolic region" in the early paper of Garett and Brattain on the MIS diode [21] (see also [22]). This particular situation is characterized by the fact that majority carriers have been repelled away from the surface, leaving a depletion charge of fixed atoms. The density of minority carriers is increased with respect to the distant bulk, but it is still negligible in the overall charge balance, and does therefore not affect the capacitance-voltage curves of the MIS diode. However, these minority carriers are the only mobile charge available at the surface. Hence, as soon as some voltage is applied between the source and the drain of a MOS transistor structure, they move by diffusion, thereby producing a drain current.

This current was ignored for years, since it was at the useless sub-microampere level, even for rather wide transistors. Indeed, MOS transistors were used, and are still mostly used, with a strongly inverted channel, i.e with a density of inversion charge comparable to, or larger than, that of the depletion charge.

The very first application that needed to limit the power consumption of integrated circuits at the microwatt level was the electronic watch. Early developments at CEH (Watchmakers' Electronic Center, Switzerland) started in bipolar technology [23], but it soon became obvious that the newly proposed CMOS technology [24] was ideally suited. The main problem in developing the logic part of the system (dominated by frequency dividers) was to obtain threshold voltages below 1 volt to ensure strong inversion at the low supply voltage of 1.3 volt in order to reach the required speed. But the heart of a watch is its crystal oscillator, for which a sufficient continuous transconductance must be created with the fraction of microampere of available current.

2 Origins of Weak Inversion (or Sub-threshold) Circuit Design

The characteristics of MOS transistors were then measured at this very low current level, and they showed the unusual exponential dependency of the drain current on the gate voltage depicted in Figure 2.1. Weak inversion then came to the attention of the digital design community under the name "sub-threshold current". Indeed, it was the residual "leakage current" that kept flowing through a MOS transistor when it was supposed to be blocked by imposing $V_{GS} = 0$.

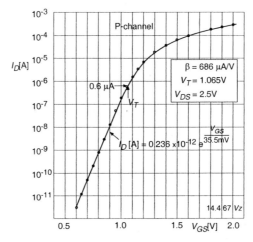

Fig. 2.1. Early measurement of the $I_D(V_{GS})$ characteristics of a P-channel metal-gate MOS transistor (cleaned-up plot from E. Vittoz' notebook, CEH, 1967).

This sub-threshold current was modeled in a succession of papers. In 1972, Barron [25] showed that this current was exponentially dependent on the surface potential (as can be expected from the same dependency of the surface charge), but he had no simple relationship with the gate voltage.

Immediately after, Swanson and Meindl [1] explained that the surface potential is related to the gate voltage by a capacitive divider, the ratio of which is approximately constant for a wide range of current. They characterized this ratio by a factor n that became the slope factor in subsequent models. In this pioneering paper, the authors applied their model to find the transfer characteristics of a CMOS inverter in weak inversion, showing that CMOS logic circuits can operate at a supply voltage as low as $8kT/q$. It is the very first publication on weak in version (or sub-threshold) digital circuits.

In 1973, Troutman and Chakravarti [26] introduced the effect of non zero source voltage and extended the model to include short-channel effects. Troutman later derived an explicit expression of the sub-threshold slope [27]. The model presented in 1974 by Masuhara, Etoh and Nagata [28] used several equations to describe the current in the whole range of operation, but the relation between gate voltage and surface potential remained complicated.

A limited small-signal model was published in 1976 with a proposal to apply it to amplification [29], but the first analog experimental circuits exploiting weak inversion were presented the same year at the second European Solid-State Circuits Conference (ESSCIRC) [30]. It is interesting to mention that one comment from the audience was that such circuits, which use the "leakage current" of transistors, could not be reliable. Among these circuits was an amplitude-regulated crystal oscillator that has since been integrated by the tens of millions in electronic watches. Another was a current reference circuit that was directly inspired from its known bipolar version. The compact model used to describe the drain current

$$I_D = I_{D0} \exp \frac{V_G}{nU_T} \left(\exp \frac{-V_S}{U_T} - \exp \frac{-V_D}{U_T} \right) \qquad (2.1)$$

was inspired from all previous publications and was derived in the extended paper published the following year [31]. In this model, all voltages are referred to the (local) substrate in order to preserve the intrinsic symmetry of the device. As was pointed out later [32], this model is very similar to the Ebers-Moll model [33] for a bipolar transistor having negligible base current. A related small signal AC model was first presented in 1977 [34]. It included the experimental evidence that channel noise is indeed the shot noise that can be expected for a barrier-controlled device.

During the following decade the interest for exploiting weak inversion in analog circuits was very limited world-wide [35, 36], and mostly concentrated in Switzerland [37, 38, 32, 39, 40, 41, 42, 43, 44, 45, 46] for the development of micropower circuits used in a variety of portable systems. These circuits had to use combinations of transistors biased at various current levels from weak to strong inversion. This was the motivation for the development of a continuous model, which was started by Oguey and Cserveny [47, 48] and became the EKV model [49].

In the late 80's, a new wave of interest for weak inversion (or sub-threshold) circuits was triggered by Mead at Caltech. He promoted them as the best way to implement analog VLSI systems that mimic the operation of the brain [50]. Mead and Maher also introduced the charge-potential linearization in a charge-based model [51], which was later adopted for the EKV model [52].

Sub-threshold digital circuits remained totally ignored until the mid-90's, with the newly recognized need of limiting the power consumption, in particular for portable systems. With modern deep submicron processes, it should be possible to reach clock frequencies much beyond 10 MHz with a supply voltage of just a few hundred millivolts and to drastically reduce thereby the power-delay product [53].

The reduction of supply voltage imposed by scaling-down process dimensions implies a reduction of the saturation voltage of transistors used in analog circuits. This is only possible by reducing the amount of channel inversion, thereby entering moderate inversion, or even weak inversion for supply voltages below half a volt.

3
Survey of Low-voltage Implementations

3.1 Technology Scaling

During the 1970s the MOSFET started gaining in popularity particularly for its dominance in low power and speed. In a retrospective paper about the development of semiconductors in the 1970s [54], Sze plotted the trend of papers published on both bipolar devices (thyristor, bipolar transistor, etc.) and unipolar devices (MOSFET, CCD, MESFET, etc.). The plot showed the number of papers on the MOSFET increasing by 20X over a decade. Sze also noted the exponential increase of MOS devices on a chip since 1959. In particular the amount of MOS memory was doubling at a faster rate than logic gates. Figure 3.1 shows the transistor scaling with time. This plot is compiled based on CMOS device technologies published in IEEE papers since 1980. Overlayed is the ITRS roadmap prediction of device scaling out to 2010 [55]. The figure shows the 3 year delay between process nodes accelerate to a 2 year development cycle.

Ever since the first integrated circuit, people have predicted how the feature-size will scale with time. In a forty-year retrospective, Svensson showed the predictions people made over time on the minimum feature-size [56]. The predictions cited different sources of failure such as lithography error, dopant fluctuations, cosmic radiation, and Drain-Induced Barrier Lowering (DIBL). Over 40 years, predictions ranged from $2\mu m$ limit in 1962 to 10nm in 2001 and even showed a prediction of 2nm by Swanson in 1960 [57].

Figure 3.2 shows the rated supply voltage defined for each process technology node defined in Figure 3.1. The published values are compared to the ITRS roadmap values. Until the sub 1-micron process node, the supply voltage remained at 5V.

In 1974, Dennard proposed the Constant Electric Field (CE) theory, where the supply voltage is scaled to keep the electric fields constant [58]. The CE theory said that for a device scaling factor of S, $(S > 1)$, as the dimensions scale with $1/S$, the drain supply voltage scales by $1/S$ to keep the electric field constant. The benefits of CE are that, as the delay decreases by $1/S$,

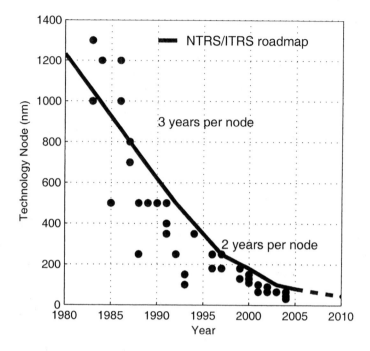

Fig. 3.1. Published transistor scaling

the power density is constant, the packing density is S^2 and the power-delay product is $1/S^3$.

Although the benefits of scaling voltage were theorized, it did not occur in practice until the 0.8μm and even 0.5μm process node. There are exceptions, e.g. IBM shown by the diamonds in Figure 3.2, where voltage scaling occurred at the 1μm node. Figure 3.2 shows where IBM process technologies are defined in relation to other technologies. Most technologies were adhering to the Constant Voltage (CV) theory, where the voltage supply remained at 5V for many generations to facilitate migrating designs and to adhere to standards. CV theory does have it's benefits: Delay scales as $1/S^2$, packing density is S^2, and power-delay is $1/S$. However, the power-density went as S^3. As the number of devices on chip grew exponentially, the CV theory exacerbated the power-density problem, and eventually voltage scaling was adopted at the sub 1-micron nodes. In retrospect, the Quasi-Constant Voltage (QCV) scaling was observed. QCV combines CE and CV and has different supply voltages for different dimensions [59][60].

Soon enough, power was not the only driver for lowering voltage with each process shrink. Device reliability also drove toward lowered voltage. Device reliability was compromised by gate oxide breakdown and hot carrier effects due to large voltages applied to the MOSFET. In [61], the authors provide an

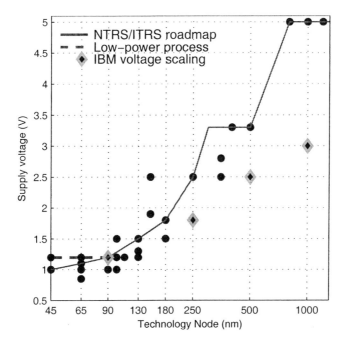

Fig. 3.2. Supply voltage vs. technology node.

equation for supply voltage that also accounts for reliability. The equation was a function of design rules and predicted 3.05V voltage for 0.5μm technology.

As technology scaling continued from 0.8μm down towards 0.25μm technologies, voltage scaled from 5V towards 2.5V, and most design migrations to new technologies did not require much CMOS circuit redesign. However, voltage scaling below 5V led to an array of challenges at the system level and new circuit implementations followed. These challenges are elaborated on in Section 3.2.

In deep sub-micron technologies, voltage scaling slowed and flattened out (Figure 3.2). Also, starting at around the 130nm node, process designers began to distinguish between high-performance processes and low-power processes. This divergence of process design trends is evident in Figure 3.2. The high-performance process tends to follow traditional voltage scaling and threshold voltage scaling. The supply voltage for 65nm high-performance processes is 1.0V, because very low threshold voltages are targeted and high-performance products can tolerate high leakage power. The low-power processes tend to target higher threshold voltage for low-leakage conditions. For high-performance modules in a low-leakage process, the rated voltage for 65nm low power processes is 1.2V.

3.2 Low-voltage Logic Designs

Voltage Scaling in Strong Inversion

The tradeoff between power and delay has been a consideration for circuit designers since the earliest integrated circuits were built. For well-designed circuits, higher speed comes at the expense of more energy consumption, and lower energy operation requires larger delays. Figure 3.3 shows the operating voltage as a function of year. This figure was compiled based on publications in the International Solid-State Circuits Conference and Journal of Solid-State Circuits. The high speed processors tended to stay with the International Technology Roadmap for Semiconductors (ITRS) predictions for voltage and technology [55].

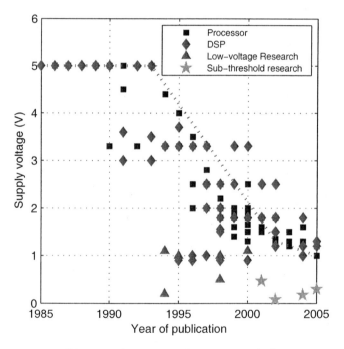

Fig. 3.3. Operating voltage vs. Year[55]

In the 1980s, Integrated Circuits (ICs) switched from NMOS to Complementary MOSFET (CMOS) designs. The dramatic improvement in power of CMOS logic relative to the NMOS style logic provided enough flexibility to allow designers to focus less attention on energy consumption.

In the early 1990s, many designs switched from BiCMOS to CMOS and started operating at voltages lower than 3V. BiCMOS designers saw that it was becoming difficult to reduce the voltage and achieve high performance designs. Some designs found an advantage in combining BiCMOS and CMOS, but soon high speed designs exclusively used CMOS processes.

In the 1990s, CMOS ICs started making the shift from traditional 5V systems down to 3.3V. In [62], the authors describe the system and design challenges in migrating a design from 5V at 1.5μm to 3.3V 1.0μm process. They cited the need for reduced power dissipation and improved device reliability. One main challenge for scaling voltage was interfacing to legacy 5V chipset requirements. This drove the design to incorporate both 5V I/O supply and a 3V internal voltage. The authors also cited many advantages in migrating to 1.5μm, such as being able to efficiently add more memory to their chip. Also they reduced the threshold voltage and achieved 75% better performance node-to-node.

In 1991, several publications featured digital signal processors (DSPs) that operated at the lower 3V range at the 0.8μm node and above, and accelerated voltage scaling relative to previous trends. The authors realized that 5V performance requirements could be relaxed for their application. Therefore, they tailored the voltage to the necessary performance requirement and were able to operate at lower voltage and to dissipate less power.

Emphasis on high performance operation has kept most energy-delay optimizations in the strong inversion region. In 1994, various papers showcased a new push towards low-power for high-performance. Power optimization techniques included architectural solutions like pipelining and parallelism, voltage scaling in strong inversion, and sizing [63]. These approaches effectively sped up circuits and then traded off the extra speed for power savings by scaling the supply voltage. This approach provided quadratic savings in energy while keeping performance (throughput) constant and was equivalent in most cases to trimming wasted energy and pushing designs closer to the Pareto optimal point for a given delay. A second paper in 1994 describes a process with zero threshold voltages and demonstrated an encoder/decoder operating in strong inversion at 200mV [64].

In the late 1990s, DSPs tended to push the voltage envelope and demonstrated many designs at around 1V operation. In 1995, NEC demonstrated a CMOS DSP in a 0.25μm process that was targeted at 0.9V operation. They were able to demonstrate a wide range of operability between 0.5V to 2.5V with their simple DSP composed of a multiplier, adder, SRAM and PLL. In 1996, NTT and Toshiba both demonstrated much more complex DSP designs at 1V and below. NTT designed a 0.9V 2-D DCT core processor for an HDTV application [65]. This dedicated chip also used body bias control of the threshold voltage and achieves 150MHz performance with 10mW power dissipation. Compared to previous implementations, this design dissipated 10% lower power. Toshiba designed a 26MOPS programmable DSP that operated at 1V [66]. This DSP was targeted for mobile applications that require extremely low power consumption but also have significant data processing workloads. In 1997, TI demonstrated a 1V programmable DSP chip that achieved 60MHz operation and dissipated 17mW for an FIR filter application [67]. The DSP was targeted for wireless application and thus required much higher throughput and extremely low power dissipation. It was able to show operation down to 0.6V.

In 1998, the University of Tokyo showed a 0.5V testchip consisting of pass transistor logic. The chip was fabricated in a $0.3\mu m$ process which typically has a 2-3V nominal supply. The process technology had threshold voltages as low as 0.1-0.2V, which normally would have extremely high standby current, but was coupled with their super-zigzag cut-off scheme that reduced the leakage from 10nA-per gate to pA-per gate in standby.

By 1999, most DSPs and processors were operating below 2V. Intel designed a Pentium 6 microprocessor for dual usage. The chip operates at 2.2V for server applications and at 1.4V for mobile operation. This was achieved by scaling the transistor channel length and migrating to CMOS technology from BiCMOS.

In 1997, [68] showed the advantages of Dynamic Voltage Scaling (DVS) for power savings. The authors showed that dynamic scaling of voltage and frequency lead to 30-50% average power savings. This paper showed that frequency scaling alone does not improve energy savings. Rather, if latency is relaxed, both voltage and frequency can scale together to produced quadratic power savings due to the relationship between active power and supply voltage. This paper also introduced the concept of voltage dithering. They showed that a good approximation to continuous voltage and frequency range is achieved with four discrete voltages.

Intel prominently showcased DVS in the low-power StrongARM processor for hand-held devices [69]. The StrongARM was able to operate at 200MHz at 0.7V, and up to 800MHz at 1.65V. Soon after many Processors and DSPs demonstrated DVS as shown in Figure 3.4.

In 2005, the concept of UDVS was introduced and demonstrated on a 90nm Kogge-Stone adder. The adder was able to operate both at 1.2V and 0.3V. Also voltage dithering was used to approximate a continuous voltage and frequency curve. This design is described in more detail in Section 9.2.

Technology scaling has now pushed power consumption back to the forefront to the point that most circuit and system designs are power-limited [70]. This trend has been evident for the last several technology generations, as designers find it increasingly important to operate their circuits on the Pareto optimal curve between energy and delay. Even though energy has become increasingly important, the majority of attention has focused on the high performance tail of the energy-delay Pareto curve. Designers tried to minimize energy consumption while meeting high performance frequency constraints.

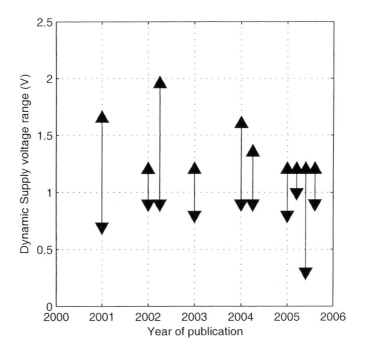

Fig. 3.4. DVS vs. year

3.3 History of Minimum Voltage

As early as 1962, Keyes published papers about the limitations of performance and power dissipation of digital circuits [71]. The paper conclued that lowering voltage is the best way to lower power dissipation. He predicted that keeping power dissipation in check would become very difficult as the transistor density started increasing. However, there were practical limits to voltage scaling due to uncertainty energy, thermal energy and the basic fact that voltage must be high enough to maintain a minimum performance. The paper concluded that the minimum possible voltage limit is not much higher than the thermal voltage ($kT/q = 0.025$V), but ultimately voltage must be above 0.5V for performance.

Then Meindl and Swanson pushed the minimum voltage lower [72]. They built a rough model of the inverter using resistors and showed that CMOS circuits have the best power-speed product in comparison to transistor-transistor logic (TTL) or emitter-coupled logic (ECL). The authors showed that there was no limit to voltage scaling as long as large delays were tolerable. This analysis was indeed true in 1971, when leakage currents were still very small. We will show in Section 6.1 that minimizing power dissipation is equivalent to minimizing supply voltage for circuits whose switching energy dominates leakage energy.

In 1972, Swanson and Meindl revised their analysis to account for the static or "off" current [1]. They devised a charge based model for the inverter in weak, strong and "weak+strong mixed" inversion. Previous work had modeled both weak and strong inversion current, but usually there was a discontinuity in the model where the two regions met. Swanson used this model to analyze the Voltage Transfer Characteristic (VTC) of the inverter and showed that the inverter operation could be simulated down to 100mV. Figure 3.5 shows the VTC published by Swanson and Meindl in 1972. To find the minimum voltage he equated the *off* current for nMOS and pMOS and calculated the inverter gain in sub-threshold. Since an inverter must have sufficient gain at $V_{DD}/2$, the minimum voltage was estimated to be $8kT/q$ or 0.2V based on device measurements for that time. An implementation of a ring oscillator was shown to work at 100mV soon thereafter [73].

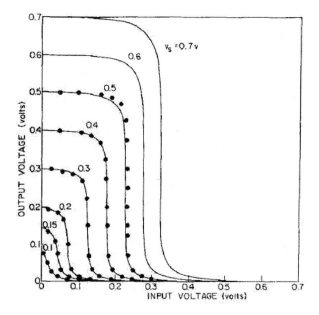

Fig. 3.5. Inverter VTC showing operation down to 100mV published by Swanson and Meindl in 1972 [1]. (© 1972 IEEE)

In 1981, Keyes analyzed in detail the limitations of digital circuits according to information theory techniques [74]. Information theory states that the minimum energy to transmit a bit is constrained by the thermal limit or thermal noise. Thus a few hundred times kT is required to electrically transmit one bit of information. Another requirement specific to digital circuits is output stability. This requirement stipulates that the output must be a non-linear function of the input. Low voltage can change the non-linearity function for many circuits. The author finally concludes that the main reason why high voltages, much higher than the thermal limit of kT/q, will be required is due to V_T variation. Threshold voltage will vary due to processing differences (e.g., dopant variation, lithography, etc.) and due to environment (e.g., temperature). Therefore, to properly switch the transistor, the supply voltage must be set higher than the threshold voltage plus margin accounting for all possible variation. Again, an implicit assumption the author makes is that strong inversion is necessary to maintain performance requirements. However, Keyes does mention that with a superconductive material or at ultra low temperature conditions, the supply voltage can be as low as 0.4V.

In 1991, Burr started investigating minimum voltage operation and also started theorizing on the optimal supply and threshold voltage pair [75]. Trends indicated that threshold voltage should decrease at the same rate as supply voltage for performance. However, increasing leakage currents place a lower bound on threshold voltage scaling. Thus, for a given process technology, the threshold voltage is fixed based on these constraints. Burr provides an equation for the user to find the minimum V_{DD} for their specific application:

$$V_{DD} = V_T + \sqrt{2 * I_{ds}/k} \qquad (3.1)$$

where k is a process dependent term, and I_{ds} is the desired I_{on} or drain current. For example, for 2μm CMOS and V_T=160mV, the voltage required is 300mV.

In a future paper, the Stanford Ultra Low Power (ULP) CMOS process technology was re-designed to use near zero V_T devices and body biasing to tune V_T up to around $V_{DD}/3$. The resulting circuits can operate down to below 100mV, but they operate the transistors in strong inversion [76].

While Burr does advocate for operating at the minimum supply voltage, he does point out many barriers to achieving this voltage. Among those barriers are level-shifting to I/O and analog supply voltages which cannot scale and are fixed by system-level standards. Another barrier is the need for internal and external noise isolation. Internal noise may be decreased using decoupling capacitors. However, unpredictable noise such as power supply variation or coupling from other sections of the chip which operate at high voltage, must be shielded from the low-voltage operating digital logic. Threshold variation again is noted as a barrier. As variation gets worse with deep sub-micron technology, significant voltage margin is needed. Burr suggests that well-biasing may solve the variation problem. The final barrier to low-voltage operation is the leakage current that increases the power-delay product at the minimum supply voltage. Using high V_T devices is suggested to lower leakage, but ultimately the optimal supply operation is higher that the minimum supply due to leakage currents. Burr also expounds on the optimal supply voltage operation that minimizes power which we discuss in Section 3.4.

20   3 Survey of Low-voltage Implementations

In 1996, Schrom provides analytical lower bounds on voltage scaling that incorporates aspects of noise margins, voltage gain and nMOS/pMOS asymmetry [77]. To find the lower bound, first the user must set their desired conditions before calculating the minimum voltage for their circuit. The paper shows the minimum voltage for various circuits range as low as 36mV for an inverter to 83mV for standard logic to 238mV for dynamic logic. The authors also propose process technology changes for low-voltage operation that scale better into low-voltage because the process is able to avoid many constraints that high-performance designs require. The author concludes that ultimately, to optimize for low-voltage, independent threshold voltage adjustment is required.

In 2001, another minimum voltage operation theory emerged [78]. To achieve the lowest possible voltage, the nMOS and pMOS off currents must be equalized. The proposed ideal limit was

$$V_{DD} = 2nkT/q \approx 57mV \qquad (3.2)$$

In an inverter test circuit, the authors use feedback to control the voltage to the wells in order to match the nMOS and pMOS current. The circuit was fabricated in a 180nm process which has a nominal voltage of 1.5V. 70mV inverter operation was demonstrated with a threshold voltage of zero volts. The authors showed 200mV operation lower energy than the minimum voltage point. Again, this shows that in deep sub-micron technologies the leakage currents make minimum voltage operation less energy efficient.

In 2002, Ono derived another minimum voltage limit by equating the nMOS and pMOS threshold voltages [79]. Using device models, they showed that the minimum V_{DD} can be reduced on an SRAM bit by 0.15-0.3V by biasing the nMOS and pMOS wells so that their V_T matched. Kao implemented a multiply-accumulate test chip designed in a triple-well process with a nominal V_T of 0.05V. The results showed that adaptive V_{DD} and V_T minimizes active power dissipation [80]. By tuning the threshold voltage, the test chip achieved operation down to 175mV.

3.4 History of Minimum Energy

For many years, minimizing voltage was thought to be the same as minimizing power consumption, because for many technology nodes static leakage current was insignificant and not considered in the analysis. In 1991, Burr did mention that the off current would limit the theoretical minimum supply voltage [81]. In the paper, he shows the optimum supply voltage depending on threshold voltage, supply voltage, logic depth, and area. The paper contains an energy curve showing the tradeoff between static energy, switching energy, activity factor and logic depth resulting in an optimum voltage where total energy is minimized. This minimum energy curve is modeled and analyzed in detail in Chapter 4.

3.4 History of Minimum Energy

Nowak later theorized in 1993 that the optimal V_{DD} possible was 0.5V for 100nm technologies but with a delay penalty. However, operation at the minimum voltage was rarely pursued because of the degradation in performance. For maximal performance, the optimal voltage was predicted to be 0.8V in 1993.

In 1997, Gonzalez used simple leakage and active current models to plot energy-delay product (EDP) versus supply voltage and threshold voltage [82]. EDP is a metric that is commonly used for servers and high-performance processors and gives more weighting to performance. The authors show for a 0.25μm technology the optimal operating point that minimizes EDP was V_{DD}=250mV and V_T=120mV. If the minimum EDP curve is steep, then the pay-off for operating at the minimum point is the greatest. However if the curve is shallow, then there is little benefit to operating at the minimum EDP. Instead, the application can achieve higher performance with little energy cost. The paper also looked at variation and its effect on the minimum EDP point. Variation increases the minimum EDP point to a higher threshold and supply voltage and flattens out the EDP curves. In Chapter 4, we extend this analysis by splitting out energy and performance into two curves, with the main goal to minimize energy instead of EDP.

In 1999, Stan gives intuition into optimal voltages and sizing for minimizing EDP [83]. He shows that EDP is the best metric to optimize because it strives to minimize power without sacrificing too much performance. A rule-of-thumb proposes the supply voltage should be slightly larger than twice threshold voltage for optimal EDP.

In 2000, Nose gives an analytical expression for the optimum supply voltage and threshold voltage point that minimizes power at a given performance [84]. Fitted technology parameters and modeling show the optimal operating point for different logical depths and activity factors. For a high performance requirement, the optimum threshold voltage ranges from 0-0.1V and optimum supply voltage follows the Semiconductor Industry Association (SIA) roadmap.

Bhavnagarwala extends the transregional model using the National Technology Roadmap for Semiconductors (NTRS) roadmap and predicts the optimal supply voltage that minimizes power to be 510mV for the 50nm technology node [85]. The authors also predict that datapaths that exploit parallelism will help to further scale the supply voltage down with new process technologies.

Measurements of a test chip with adaptive supply and body bias display a minimum power point for a given performance and show how forward-biased diode currents (from body biasing) can make the theoretical optimum unreachable [80]. More recently, derivations of the sensitivities of energy and delay to different parameters support a more formal methodology for building optimum energy circuits [86].

3.5 Survey of Sub-threshold CMOS Circuits

Several papers in Section 3.3 agree that either higher supply voltage or near zero threshold voltage are necessary to minimize energy dissipation without unreasonably large leakage current or delay penalties [78][76]. Sub-threshold circuits are ideal for applications where performance is not critical but minimizing energy consumption is key.

An ideal application for ultra-low voltage CMOS circuits was the wristwatch [87]. The wristwatch did not need very high speed circuits and thus could sacrifice performance to lower power. In 1972, Vittoz et al. designed a CMOS frequency divider that operated from a 1.35V mercury battery with 2MHz performance. The authors did not push the voltage to weak inversion because the maximum performance was below 1MHz with the 6-10 μm channel lengths. Additionally, at 1.35V the digital power was extremely small when compared to the battery leakage and display power. The process pushed the threshold voltages to less than 1.1V for the nMOS and greater than -0.45V for the pMOS. In the paper the authors did show sub-threshold operation as low as 0.95V, which was a breakthrough result. [88]

Other than this demonstration, most sub-threshold circuit design focused on sub-threshold analog circuits. Vittoz demonstrated analog circuits such as an amplitude detector, a quartz ring oscillator, a bandpass amplifier, etc. operating in weak-inversion [31]. Lyon and Mead designed many transconductance amplifiers in sub-threshold for cochlea applications [89].

Sub-threshold CMOS circuits emerged again in 2001 as a solution for a hearing-aid application. The hearing-aid required very low frequency clocks, and thus could operate in sub-threshold. Various CMOS datapath circuits were designed such as a Delayed Least Mean Square (DLMS) filter [90] and adders [91]. The adder was built in 0.35μm technology and operated as low as 0.47V.

Also, this research explored different logic families for sub-threshold. The adder design used variable threshold voltage CMOS logic that has a controller that biases the pMOS and nMOS backgates. Another logic family explored was Pseudo-NMOS (P-nMOS) [90], where a weak pull-up pMOS is always on. A third logic family explored was dynamic threshold voltage CMOS, where the backgates of each transistor is connected to its own gate. We provide an additional analysis of sub-threshold logic styles in Section 6.3.

In 2002, Deen demonstrated an ultra-low power voltage-controlled oscillator (VCO) design using ring oscillators [92]. The VCO is able to operate down to 80mV, which is significantly lower than the nominal voltage of 1.8V for a 180nm design. Body biasing is used to tune the speed of the VCO as it transitions from strong to weak inversion.

This book discusses sub-threshold designs demonstrated in 2003 and later. In 2004, a sub-threshold FFT was implemented in 180nm bulk CMOS and operated at 180mV [2]. The FFT uses a sub-threshold digital logic library and a register file design for ultra-low voltage operation. Design techniques for sub-threshold logic are described in Chapter 6, and register file design considerations appear in Section 7.1. In 2005, a sub-threshold SRAM was implemented in 65nm bulk CMOS and operates to below 400mV [93][3]. More sub-threshold SRAM design considerations are described in Section 7.2.

4
Minimizing Energy Consumption

This chapter introduces the concept of a minimum energy operating point for a digital design. When the circuit operates at this minimum energy point, it consumes less energy per operation than at any other point in the parameter space. While many different parameters can impact the minimum energy point, this chapter focuses on the dual knobs of power supply and threshold voltage. Energy and performance simulations of a variable activity factor characterization circuit over a range of supply voltages and threshold voltages provide insight into the behavior of the minimum energy point. After analyzing the results of these simulations, we derive analytical expressions for the optimum V_{DD} and V_T values at the minimum energy point from well-known equations for sub-threshold current. This model provides a base for analyzing how variables such as workload, duty cycle and temperature affect the minimum energy point. Measurements of a test chip that implements a programmable Finite Impulse Response (FIR) filter confirm the results from the minimum energy point models.

4.1 Energy-Performance Contours

By examining the energy and delay contours for a simple circuit over supply voltage (V_{DD}) and threshold voltage (V_T) we show that minimum energy operation occurs in the sub-threshold operation region. This optimum point changes with different activity factors and with threshold variation [94].

4.1.1 Variable Activity Factor Circuit

Simulations of a variable activity factor ring oscillator circuit demonstrate the optimal V_{DD} and V_T for CMOS circuits. The circuit consists of an 11-stage 2-input NAND ring oscillator (RO) (Figure 4.1). The stage depth of the RO is chosen to emulate the logic depth of a processor pipeline. We define the performance as the simulated ring oscillator's frequency.

26   4 Minimizing Energy Consumption

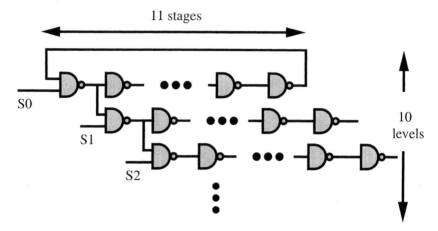

Fig. 4.1. Variable activity factor circuit (© 2002 IEEE)

A variable activity factor is achieved by placing nine additional delay chains in parallel which are driven by the RO and run at the same frequency. The delay chains are enabled/disabled using selector inputs to vary the activity factor. When all of the chains are activated the activity factor is 1, since all nodes in the ring oscillator switch during each cycle. By asserting inputs S_0 to S_4 and deasserting signals S_5 to S_9, the activity factor is 0.5, because only half of the levels are switching. The variable activity factor circuit allows for emulation of a wide range of circuits because logic gates typically have activity factors much lower than 1.

As activity factor decreases, the amount of switching also decreases. This causes leakage energy to becomes more significant in the overall power dissipation.

4.1.2 Energy-Performance Contours

Simulations of the energy-performance variable activity factor ring oscillator circuit in a 0.18μm process are shown in Figure 4.2. In these simulations, V_{DD} varies from 0.1V to 0.6V, and the threshold voltage for both nMOS and pMOS varies from 0V to 0.6V. The activity factor is set to 1. The combination of V_{DD} and V_T for which the circuit consumes the least amount of energy is marked by the large filled circle and occurs at V_{DDopt}=130mV and V_{Topt}=370mV. Contours showing a constant average energy per cycle trace rough circles around the minimum energy point. The values associated with the energy contours are normalized to the minimum energy point.

Since this operating point is significantly below the typical voltages for a 0.18μm process [(V_{DD},V_T)=(1.8V,0.45V) nominally], the RO switches at a very low clock frequency of 75kHz. Figure 4.2 confirms that the optimal operating point for this circuit falls in the sub-threshold region.

4.1 Energy-Performance Contours

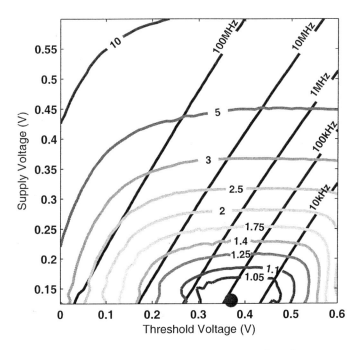

Fig. 4.2. Minimum energy point and constant energy and performance contours of the variable activity factor circuit with activity equal to one (© 2002 IEEE)

Figure 4.2 also shows contour lines marking constant performance of the circuit. The ring oscillator frequency varies between 10kHz and 500MHz.

The occurrence of a minimum energy point is due to the relationship between energy and latency at the given V_{DD} and V_T. At strong inversion supply voltages, switching energy dominates the total power. As V_{DD} is lowered from 1V to 100mV, a dramatic reduction of switching energy occurs because switching energy scales quadratically with supply voltage:

$$E_{DYN} = \alpha C V_{DD}^2 \qquad (4.1)$$

However, as the supply is reduced to sub-threshold voltage levels, the propagation delay increases exponentially. This leads to a corresponding exponential increase in leakage energy since the leakage current is accumulated over a longer period of time.

Figure 4.3 shows the relationship between switching energy and leakage energy in the characterization circuit as a function of V_{DD} with V_T fixed at 500mV and an activity of one. Mathematically, the minimum energy point occurs where the slopes of the leakage energy and active energy are equal in magnitude and opposite in sign. The figure shows that the minimum energy point for this example occurs in the sub-threshold region at $V_{DD} = 150$mV.

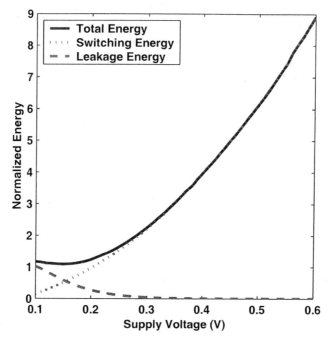

Fig. 4.3. Minimum energy point for a fixed threshold voltage

4.1.3 Activity Factor

Figure 4.4 illustrates the impact of activity factor on the minimum energy point. By using the selector inputs, the activity factor is decreased to $\alpha=0.5$ in Figure 4.4 (a) and to $\alpha=0.1$ in Figure 4.4 (b). These energy contours show that as activity factor decreases, the V_{DD} and V_T associated with the minimum energy operating point increase. Decreased activity factor leads to an overall decrease of switching energy, while leakage energy remains constant with activity factor. This causes leakage energy to contribute relatively more to the total energy.

Analysis of these contours provides us with a better understanding of how to minimize energy consumption in system design. All applications can be categorized into one of two cases.

Case 1: Processing speed is not critical. For this case, the optimal operating point occurs at the minimum energy contour. Circuits should always operate at the optimal supply and threshold voltage because, even though the performance is low, it is still sufficient for the application. When the task is finished, then the processor should enter a power down state or switch to a low-leakage mode to prevent further leakage energy dissipation.

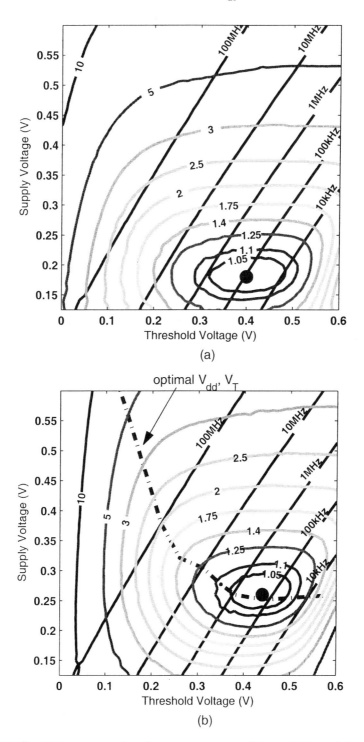

Fig. 4.4. Constant energy and performance contours of the variable activity factor

Case 2: Processing speed is critical. For this case, an optimal operating curve is extracted from the contours which contains those points where one performance contour is tangent to one energy contour. These points are given by the dotted line in Figure 4.4 (b). This is called the optimal operating curve because it shows the V_{DD} and V_T which minimizes energy as a function of frequency. The system should set its V_{DD} and V_T along this curve for minimum energy dissipation. If the frequency required is less than the frequency of the optimal energy contour then this case reverts back to Case 1.

4.2 Modeling Minimum Energy Consumption

Although the energy contours in the previous section provide good insight related to minimum energy operation, still a specific analysis of the minimum energy point is needed. This section describes an analytical model that supplies such an analysis. First, this section introduces a model of the MOSFET drain current in the sub-threshold region. The current model serves as a basis for the minimum energy point analysis. The model that we develop for the minimum energy point allows quick estimation of important trends and the impact of key parameters on the system energy.

4.2.1 Sub-Threshold Leakage Current Models

In sub-threshold operation, the channel of the transistors is not inverted and current flows by diffusion. Equation (4.2) is a basic equation for modeling sub-threshold current and total off current.

$$I_{D:sub-threshold} = I_o \exp\left(\frac{V_{GS} - V_T}{nV_{th}}\right) \quad (4.2)$$

Equation (4.3) shows the same basic equation with low V_{DS} roll-off: [49][95][96]:

$$I_{D:sub-threshold} = I_o \exp\left(\frac{V_{GS} - V_T}{nV_{th}}\right)\left(1 - \exp\left(\frac{-V_{DS}}{V_{th}}\right)\right) \quad (4.3)$$

where I_o is the drain current when $V_{GS} = V_T$ given in (4.4) [49][96].

$$I_o = \mu_o C_{ox} \frac{W}{L}(n-1)V_{th}^2 \quad (4.4)$$

For the model that we present in this chapter, we assume that total drain current in sub-threshold equals sub-threshold current. Section 4.2.2 comments on the validity of this assumption. As expected for diffusion current, (4.3) shows that I_D depends exponentially on V_{GS}. V_T is the transistor threshold voltage, n is the sub-threshold slope factor ($n = 1 + C_d/C_{ox}$), and V_{th} is the thermal voltage, $V_{th} = kT/q$. The parenthetical term on the right models the roll-off in current that occurs when V_{DS} drops to within a few times V_{th}.

4.2 Modeling Minimum Energy Consumption

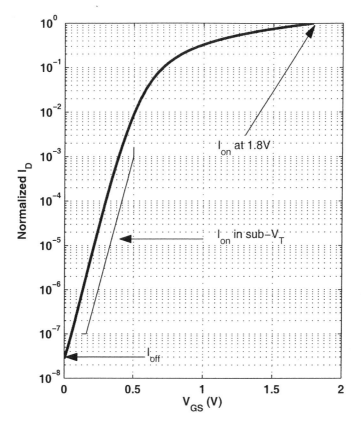

Fig. 4.5. MOSFET drain current, I_D, versus gate to source voltage, V_{GS} in 0.18μm with $V_{DD} = 1.8$V. In sub-threshold, I_D varies exponentially with V_{GS}. In fact, we define V_T by looking at where the I_D curve deviates from its original exponential trajectory. In the sub-threshold region, the I_{on}/I_{off} ratio reduces relative to strong inversion.

To model Drain-Induced Barrier Lowering (DIBL), (4.3) can include a linearized DIBL coefficient, η, as in [95]:

$$I_D = I_o \exp\left(\frac{V_{GS} - V_T + \eta V_{DS}}{nV_{th}}\right)\left(1 - \exp\left(\frac{-V_{DS}}{V_{th}}\right)\right) \quad (4.5)$$

The sub-threshold slope of the transistor is defined as:

$$S = nV_{th} \ln 10 \quad (4.6)$$

S gives the inverse of the slope of I_D versus V_{GS} in millivolts per decade of change in I_D. The ideal value for S at room temperature is 60mV/decade,

32 4 Minimizing Energy Consumption

and it occurs at the limit when $n = 1$. Plugging (4.6) into (4.5) gives the sub-threshold drain current in a different form:

$$I_D = I_o 10^{\left(\frac{V_{GS}-V_T+\eta V_{DS}}{S}\right)} \left(1 - \exp\left(\frac{-V_{DS}}{V_{th}}\right)\right) \qquad (4.7)$$

Figure 4.5 shows the drain current of a MOSFET versus its gate to source voltage, V_{GS}, across the full range from 0V to 1.8V, which is V_{DD} for this 0.18μm technology. At low values of V_{GS} in the sub-threshold region, I_D varies exponentially with V_{GS} as expected. We define the threshold voltage, V_T, by the point on the I_D versus V_{GS} plot where I_D ceases to depend exponentially on V_{GS}. This point occurs at around half a volt for the transistor in the figure.

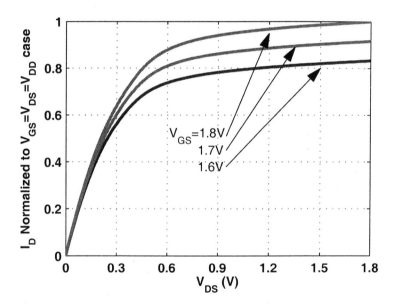

Fig. 4.6. I_D versus V_{DS} curves for three values of V_{GS} in a 0.18μm process with $V_{DD} = 1.8$V. This above-threshold inverter shows velocity saturation, since I_D increases linearly with V_{GS}.

Figure 4.6 shows standard I_D versus V_{DS} curves for strong inversion. The current in this plot clearly demonstrates the linear region and the velocity saturation region (not saturation, because I_D changes linearly with V_{GS} instead of quadratically). Figure 4.7 shows the same curves in sub-threshold. The curves show the exponential dependence on V_{GS}, but they otherwise appear quite similar to the strong inversion curves in their shape. The 'quasi-linear' region comes from the roll-off of current at low V_{DS}, as seen in (4.3). Unlike strong inversion, the onset of this roll-off depends only on V_{DS} and not on

4.2 Modeling Minimum Energy Consumption

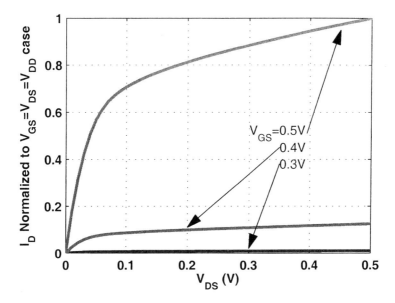

Fig. 4.7. I_D versus V_{DS} curves for three values of V_{GS} in a 0.18μm process, but $V_{DD} = 500$mV, so the inverter is in the sub-threshold region. I_D increases exponentially with V_{GS}.

V_{GS}. In strong inversion, the V_{DS} dependence in the velocity saturation region results from channel length modulation and is commonly modeled with the Early voltage. In sub-threshold, the V_{DS} dependence in the 'quasi-saturation' region results from DIBL and can be modeled with the DIBL coefficient as in (4.5) or (4.7).

Other models extend the basic form of the sub-threshold current equation to account for the transitional region between weak and strong inversion. The model in [84] uses a piecewise fitted expression to smooth out the discontinuity between the regions. A complicated piecewise current model in [85] uses physical parameters to account for current across all regions of operation. The EKV model, which is covered in Chapter 5, has become widely used especially in the context of low-voltage analog and digital circuits [49].

For gate level circuit design or for careful analysis of analog circuits, one of the higher order models makes sense. This is especially true when the circuit operates at the edge of weak inversion such that the simple equations no longer suffice. For the purposes of modeling the minimum energy point in this chapter, however, the simple sub-threshold current modeled by equation (4.2) suffices. Two reasons make this the case. First, we assume that the minimum energy point occurs well into the sub-threshold region such that (4.2) holds. Secondly, the model is intended to give an estimation of the energy-consuming behavior of large circuits, so it uses lumped fitted parameters to simplify the calculations. The subtleties of higher order current models are not necessary for this type of estimation.

One notable shortcoming of the simple sub-threshold equation is that it does not account for changes in V_T with transistor size. In Deep Sub-Micron (DSM) technologies, various higher order effects such as the Short Channel Effect, Reverse-Short Channel Effect (e.g. [97]), Narrow Channel Effect, and Reverse Narrow Channel Effect (e.g. [98]) all cause V_T, and thus I_D, to vary for small device dimensions.

4.2.2 Other Components of Current

Sub-threshold current is not the only type of leakage current that affects deep sub-micron transistors. This section briefly describes other types of leakage current and other sub-threshold current effects and discusses their impact in the sub-threshold region.

As technologies scale, the gate oxide gets progressively thinner. The higher resulting electrical field across the oxide enables carriers to tunnel through the oxide [96]. None of the strong inversion energy optimization work cited in Section 3.4 accounts for gate leakage even though gate leakage contributes significantly to total leakage in deep sub-micron technologies. For sub-threshold analysis, though, it is reasonable to ignore gate leakage. Gate tunneling current has a very strong dependence on the voltage across the gate, so it decreases with supply voltage much more quickly than does sub-threshold current. As a result, gate leakage becomes negligible in the sub-threshold region except in rare cases.

Gate-Induced Drain Leakage (GIDL) is a leakage current that is generated at the edge of the drain and terminates in the body of the transistor. GIDL current appears for high V_{DS} values in combination with low V_{GS} [96]. On a semilog plot of I_D versus V_{GS}, GIDL appears as a "tail" where the current begins to increase when V_{GS} approaches 0 and continues to increase for negative V_{GS}. For sub-threshold operation, the lower V_{DS} reduces the electric field across the drain such that GIDL becomes negligible.

As mentioned in Section 4.2.1, DIBL occurs in short channel transistors because the depletion regions around the source and drain actually overlap slightly, lowering the source potential barrier and increasing current [96]. As V_{DS} increases, the depletion region at the drain grows, further lowering the barrier due to larger overlap with the source depletion region and increasing current. The DIBL effect still impacts current in sub-threshold, but the lower V_{DS} means that the impact is less than at strong inversion voltages.

Reverse-biased diode leakage from the source and drain to the bulk also contributes to overall leakage current. This junction leakage results from a combination of minority carrier diffusion and drift at the depletion region edge and electron-hole pair generation inside the depletion region [96]. Process technologies generally are designed to make this pn-junction leakage small relative to sub-threshold current. Since the junction leakage scales with V_{DD}

and temperature in a similar fashion to sub-threshold current, it is negligible across the full range of supply voltage.

Although these other components of current can be quite significant for normal strong inversion operation, they tend to be negligible in the sub-threshold region. One scenario where this may not be accurate is in the case of very low temperatures. Since gate leakage does not depend strongly on temperature but sub-threshold current decreases exponentially when it grows colder, gate leakage can become significant again. This is especially true if the sub-threshold current is reduced further by some other mechanism (e.g. Reverse Body Bias (RBB)). Nevertheless, except for rare cases, sub-threshold current dominates in this region of operation. This allows us to equate total current in the sub-threshold region to sub-threshold current.

4.2.3 Minimum Energy Point Model

In this section, we derive a closed form solution for the optimum V_{DD} and V_T for a given frequency and technology operating in the sub-threshold regime ($V_{DD} < V_T$). The model we develop uses fitting parameters that are normalized to a characteristic inverter in the technology of interest. Since other gates are normalized to this inverter, its size is arbitrary. However, the minimum sized inverter is a good choice for simplicity.

Equation (4.8) shows the well-known form for the delay of an inverter in above-threshold from [84].

$$t_d = \frac{KC_g V_{DD}}{(V_{DD} - V_T)^\alpha} \tag{4.8}$$

The denominator of (4.8) is the *on* current of the inverter in above-threshold. We adopt the form of (4.8) for modeling the inverter in sub-threshold. Equation (4.9) shows the propagation delay of a characteristic inverter with output capacitance C_g in sub-threshold:

$$t_d = \frac{KC_g V_{DD}}{I_{o,g} \exp\left(\frac{V_{DD} - V_{T,g}}{nV_{th}}\right)} \tag{4.9}$$

As with (4.8), K is a delay fitting parameter. The expression for current in the denominator of (4.9) models the *on* current of the characteristic inverter, so it accounts for transitions through both nMOS and pMOS devices. Thus, the terms $I_{o,g}$ and $V_{T,g}$ are fitted parameters that do not correspond exactly with the MOSFET parameters of the same name unless the nMOS and pMOS are symmetrical (note that the common expression for strong inversion delay in 4.8 assume symmetrical pMOS and nMOS). Operational frequency is simply:

$$f = \frac{1}{t_d L_{DP}} \tag{4.10}$$

where L_{DP} is the depth of the critical path in characteristic inverter delays. Thus, the total delay along the critical path of a circuit is:

$$T_D = \frac{1}{f} = t_d L_{DP} \tag{4.11}$$

The exponential dependence of current on V_T and temperature results in large variations in delay across process and temperature corners. Likewise, the total leakage current and leakage power of a circuit will vary by a few orders of magnitude with process and temperature variation. Although this large variation is a problem for certain types of systems, it is not the focus of this chapter. In a severely energy-constrained system where the frequency of operation is less important than energy conservation, energy per operation is the more important metric. The following analysis shows that the energy per operation is much less sensitive to process and temperature changes than are delay and leakage current.

Without loss of generality, we assume that one operation occurs each cycle. We also ignore leakage from times after the current cycle because it either is accounted for in the next cycle or is addressed by a shutdown mode that is optimized separately. Dynamic (E_{DYN}), leakage (E_L), and total energy (E_T) per cycle are expressed in (4.12)-(4.14) [99][100], assuming rail-to-rail swing ($V_{GS} = V_{DD}$ for *on* current).

$$E_{DYN} = C_{eff} V_{DD}^2 \tag{4.12}$$

$$\begin{aligned} E_L &= I_{leak} V_{DD} T_D \\ &= W_{eff} I_{o,g} \exp\left(\frac{-V_{T,g}}{nV_{th}}\right) V_{DD} t_d L_{DP} \\ &= W_{eff} K C_g L_{DP} V_{DD}^2 \exp\left(-\frac{V_{DD}}{nV_{th}}\right) \end{aligned} \tag{4.13}$$

$$\begin{aligned} E_T &= E_{DYN} + E_L \\ &= V_{DD}^2 \left(C_{eff} + W_{eff} K C_g L_{DP} \exp\left(-\frac{V_{DD}}{nV_{th}}\right) \right) \end{aligned} \tag{4.14}$$

Concurrent work in [101] and [53] arrive at similar equations for sub-threshold energy consumption. The authors of [101] use curve-fitting to get close estimates to the optimum V_{DD} of different logic circuits, whereas we solve analytically for the optimum point.

Equations (4.12)-(4.14) extend the expressions for current and delay of an inverter to arbitrary larger circuits. This extension sacrifices accuracy for simplicity since the fitted parameters cannot account for all of the details of every circuit. Thus, C_{eff} is the average total switched capacitance of the entire circuit, including the average activity factor over all of its nodes. Likewise, W_{eff} estimates the average total width that contributes to leakage current. L_{DP} is the logic depth of the critical path normalized to the delay of the characteristic inverter. Solving this set of equations provides a good estimate of the optimum for the average case and shows how the optimum point depends on the major parameters. Differentiating (4.14) and equating to 0 allows us to solve for V_{DDopt}. Equation (4.15) shows the derivative of the total energy with respect to V_{DD}.

$$\frac{\partial E_T}{\partial V_{DD}} = 2C_{eff}V_{DD} + \left(2 - \frac{V_{DD}}{nV_{th}}\right) W_{eff} K C_g L_{DP} V_{DD} \exp\left(\frac{-V_{DD}}{nV_{th}}\right) \quad (4.15)$$

Setting (4.15) equal to zero and rearranging the equation yields:

$$2C_{eff}V_{DD} + \left(2 - \frac{V_{DD}}{nV_{th}}\right) W_{eff} K C_g L_{DP} V_{DD} \exp\left(\frac{-V_{DD}}{nV_{th}}\right) = 0$$

$$\left(2 - \frac{V_{DD}}{nV_{th}}\right) W_{eff} K C_g L_{DP} \exp\left(\frac{-V_{DD}}{nV_{th}}\right) = -2C_{eff}$$

$$\left(2 - \frac{V_{DD}}{nV_{th}}\right) \exp\left(\frac{-V_{DD}}{nV_{th}}\right) = \frac{-2C_{eff}}{W_{eff} K C_g L_{DP}}$$

$$\left(2 - \frac{V_{DD}}{nV_{th}}\right) \exp\left(2 - \frac{V_{DD}}{nV_{th}}\right) = \frac{-2C_{eff}}{W_{eff} K C_g L_{DP}} \exp(2) \quad (4.16)$$

The analytical solution for V_{DDopt} from (4.16) is given in (4.17):

$$V_{DDopt} = nV_{th}\left(2 - \text{lambertW}\left(\frac{-2C_{eff}}{W_{eff} K C_g L_{DP}} \exp(2)\right)\right) \quad (4.17)$$

If we define the argument to the Lambert W function as β:

$$\beta = \frac{-2C_{eff}}{W_{eff} K C_g L_{DP}} \exp(2) \quad (4.18)$$

then (4.17) is subject to the constraint in (4.19):

$$\beta > -\exp(-1) \quad (4.19)$$

See [102] regarding the Lambert W function and its constraints. Now, substituting (4.9) into (4.10) gives V_{Topt} for a given f:

$$V_{Topt} = V_{DDopt} - nV_{th} \ln\left(\frac{fKC_g L_{DP} V_{DDopt}}{I_{o,g}}\right) \quad (4.20)$$

If the argument to the natural log in (4.20) exceeds 1, then the assumption of sub-threshold operation no longer holds because $V_{Topt} < V_{DD}$. This

4 Minimizing Energy Consumption

constraint shows that there is a maximum achievable frequency for a given circuit in the sub-threshold region. Equations (4.17) and (4.20) give the optimum supply voltage and threshold voltage for sub-threshold circuits consuming the minimum energy for a given frequency.

Assuming a standard technology where V_T is fixed (i.e. - no triple wells for body biasing), the problem becomes finding the optimum V_{DD} and frequency to minimize energy for a given design. The optimum V_{DD} for minimizing energy per cycle in this scenario still is given by (4.17), and the optimum frequency is given by (4.10) at $V_{DD} = V_{DDopt}$.

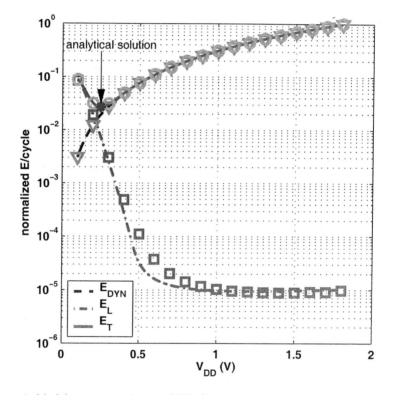

Fig. 4.8. Model versus simulation of FIR filter showing minimum energy point and contribution of active and leakage energy. Markers are simulation values, lines are model [84]. Analytical solution from eqs. (4.17) and (4.14) is shown. (© 2005 IEEE)

Figure 4.8 shows the energy profile of an 8-bit, 8-tap parallel programmable FIR filter versus V_{DD}. The contributions of active and leakage energy are both shown. The lines on the plot show the results of numerical equations using a transregional current model [84], and the markers show the simulation values. The analytical solution (small star) matches the numerical model and

4.2 Modeling Minimum Energy Consumption

simulations with less than 0.1% error. The optimum point is $V_{DDopt} = 250\text{mV}$ with a frequency of 30kHz. Equation (4.17) provided the optimum V_{DD} for the analytical solution, and substituting this value into equation (4.14) gave the total energy. Figure 4.9 shows how the delay (T_D) and current (I_{LEAK}) components of leakage energy per cycle (E_L) vary with supply voltage. As V_{DD} reduces, the current decreases due to the DIBL effect, but the delay increases exponentially in the sub-threshold region, leading to the overall increase in sub-threshold leakage energy.

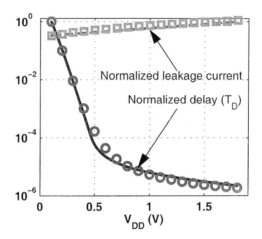

Fig. 4.9. Normalized leakage current (I_{LEAK}) and delay (T_D). Markers are simulation and lines are model [84]. Although DIBL caused I_{LEAK} to decrease, the exponential increase in T_D causes leakage energy (E_L) to increase in sub-threshold. (© 2005 IEEE)

Equation (4.17) shows that V_{DDopt} value is independent of frequency and V_T. Instead, it is set by the relative significance of dynamic and leakage energy components as expressed in equation (4.18). E_L increases compared to the characteristic inverter in two ways. First, the ratio of C_{eff}/W_{eff} decreases, indicating that a greater fraction of the total width is idle and thereby drawing static current without switching. Secondly, L_{DP} can increase. The larger resulting period gives more time for leakage currents to integrate, raising E_L. Figure 4.10 shows V_{DDopt} versus β for three examples. The first example is a sub-threshold FFT processor described in Section 9.1. The FFT has V_{DDopt} at 350mV. The second example is the FIR filter previously described which has V_{DDopt} at 250mV. The third example is a ring oscillator. The figure shows the beta and V_{DDopt} for the first two examples at 250mV and 350mV lie along the optimal curve. The ring oscillator fails to meet the constraint and thus cannot be plotted on the optimal curve. To see why, consider a single inverter

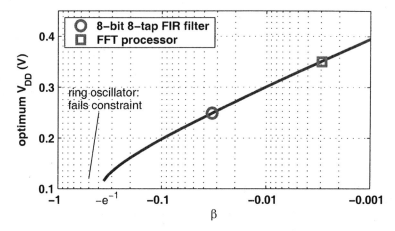

Fig. 4.10. V_{DD} optimum calculated with equation (4.17). β for ring oscillator ($L_{DP} = 21$) fails constraint. β for 8x8 parallel FIR filter and scalable FFT processor [103] also shown. (© 2005 IEEE)

with activity factor of one; W_{eff} and L_{DP} equal one, C_{eff} is close to C_g, and β does not meet the constraint in equation (4.19). Mathematically, this means that the derivative of E_T never equals zero. Physically, the leakage component for the single inverter with high activity remains insignificant compared to dynamic energy over all supply voltages, as shown in Figure 4.11. The true optimum V_{DD} in this case is the lowest voltage for which the circuit functions. Circuits with higher relative leakage energy, like the FIR filter or FFT processor, have less negative β and thus higher V_{DDopt}.

Figure 4.12 shows theoretically why V_{DDopt} is independent of V_T. As V_T decreases in the figure, the sub-threshold current increases exponentially as shown by the rise in E_L at above-V_T voltages. The sub-threshold *on* current also increases exponentially, so T_D decreases exponentially in sub-threshold and offsets the rise in I_{LEAK} such that E_L does not change in the sub-threshold region. When V_T decreases too far, then $V_{DDopt} > V_T$ so the sub-threshold equations become invalid. The figure shows that E_L physically exceeds E_{DYN} for extremely low V_T in this filter example. Of course, the advantage to lowering V_T is increased performance in the sub-threshold region for the same energy per operation.

In practice when V_T is very large, V_{DDopt} does not follow this derivation. As seen in Figure 4.4, V_{DDopt} tails off. The reason why this occurs is that for large V_T, a slow input slew rate for the circuit is caused by the long delays and short-circuit current increases. Also, as the supply voltage is lowered, the circuit does not switch rail-to-rail, and the non-ideal logic levels result in

4.2 Modeling Minimum Energy Consumption

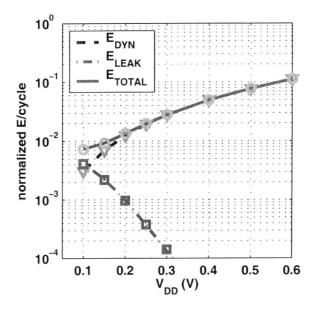

Fig. 4.11. Energy per operation versus V_{DD} for a 21-stage ring oscillator has no minimum point. Markers show simulation data and lines show equations.(© 2005 IEEE)

static current consumption. These two factors are not modeled in the energy minimization modeling and cause the V_{DDopt} to tail off.

Calibrating the model to match a generic circuit requires the fitting of only three parameters once the delay and leakage of the characteristic inverter are known. C_{eff} is the average effective switched capacitance of the entire circuit, including the average activity factor over all of its nodes, short circuit current, glitching effects, etc. To calibrate the model, C_{eff} is estimated by measuring average supply current and solving $C_{eff} = I_{avg}/(fV_{DD})$. W_{eff} estimates the average total width, relative to the characteristic inverter, that contributes to leakage current. W_{eff} is determined by simulating the circuit's steady-state leakage current and normalizing to the characteristic inverter. Since W_{eff} is a function of circuit state, averaging the circuit leakage current for simulations over many states improves the total leakage estimate. Simulating to exercise the circuit's critical path, measuring its delay, and normalizing to the characteristic inverter provides the logic depth, L_{DP}. Solving this set of equations provides a good estimate of the optimum for the average case and shows how the optimum point depends on the major parameters.

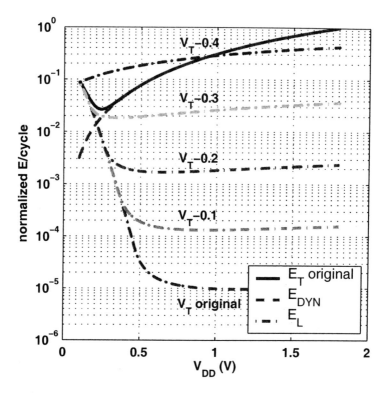

Fig. 4.12. Lowering V_T does not improve the energy per operation in the sub-threshold region, but it will increase performance at the minimum energy point.(© 2005 IEEE)

4.3 Minimum Energy Point Dependencies

The optimum V_{DD} to minimize energy will vary depending on the circuit's operating scenario, environment, and temperature. This section examines the impact of these parameters on the minimum energy point.

4.3.1 Operating Scenario

Figure 4.8 graphically confirms the trend apparent in equation (4.17). Any relative increase in the leakage component of energy per cycle will push V_{DDopt} higher, and the frequency at the optimum point also increases. In the figure, this corresponds to any decrease in E_{DYN} or increase in E_L. Likewise, any decrease in E_L or increase in E_{DYN} will lower the optimum V_{DD}. These types of changes can occur for a given circuit without changing its intrinsic attributes.

4.3 Minimum Energy Point Dependencies

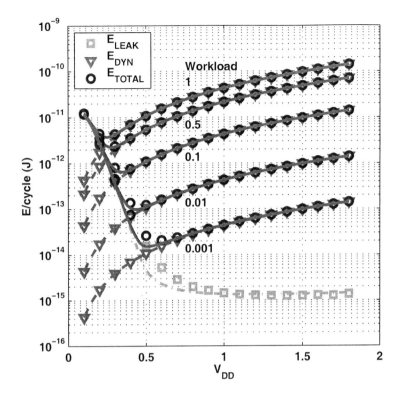

Fig. 4.13. Energy versus V_{DD} for changing workload.

For example, consider using the FIR filter in a system whose workload, ω, changes widely. This might be in a video context where the processing per frame depends on the difference between consecutive frames. If the current frame is nearly identical to the previous, then very little work is required. A scene change, on the other hand, could demand the maximum number of computations. Assuming the clock is gated when no computation is required (i.e. no mid-cycle shutdown is available) and normalizing to one cycle, C_{eff} per cycle becomes ωC_{eff} in equation (4.18).

Figure 4.13 shows the impact of a changing workload on the energy characteristics of the 8-bit, 8-tap FIR filter as an example. We vary the workload over three orders of magnitude to show the effect. The active energy decreases in proportion to workload because the amount of switched capacitance per operation has decreased. The leakage for the given operation stays constant, however, so the minimum energy point moves to lower energy and higher voltage with decreasing workload.

Duty cycle, d, also can vary widely. A lower duty cycle means that the circuit spends more idle time (e.g. waiting for data but unable to shutdown)

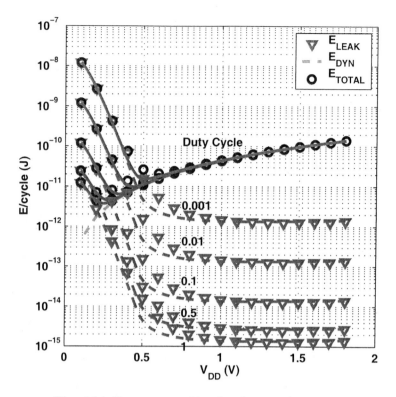

Fig. 4.14. Energy versus V_{DD} for changing duty cycle.

for each completed active operation. Consequently, the leakage contribution per operation increases, which corresponds to replacing W_{eff} with W_{eff}/d in equation (4.18). Figure 4.14 shows the impact of a changing duty cycle on the energy characteristics of the FIR filter. Again, a wide range of values is shown for the duty cycle. The longer idle time spent for each operation means that leakage energy increases, but the active energy stays constant. As with workload decreases, the minimum energy point moves to higher voltages, but the total energy increases in this case.

Normalizing to one cycle, we include duty cycle and workload in the analytical model and solve the equation set again to find the optimum V_{DD}, resulting in a new equation for β.

$$\beta = \frac{-2\omega C_{eff}}{\frac{W_{eff}}{d}KC_gL_{DP}}\exp(2) \tag{4.21}$$

Figure 4.15 and Figure 4.16 show the effects of workload and duty cycle on the minimum energy and optimum V_{DD} of the FIR filter. The figures compare the numerical result with the analytical model and with simulation.

The supply voltage for the simulations was quantized in 100mV increments. The quantization causes most of the error for values of ω and d close to one. The error in modeled energy at low values of ω and d occurs because the optimum V_{DD} has exceeded V_T. Thus the assumption of sub-threshold operation implicit to the analytical model becomes invalid. The numerical model is also less accurate in that region. The analytical result matches the numerical values quite well until V_{DDopt} nears V_T.

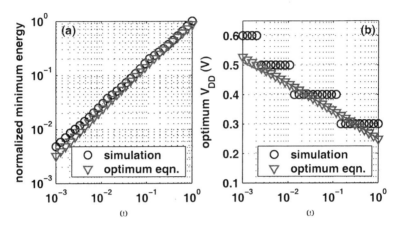

Fig. 4.15. Normalized energy (a) and V_{DDopt} (b) for FIR versus workload, ω. Simulation V_{DD} quantized to 100mV. (© 2004 IEEE)

Large reductions in either ω or d result in lower total energy per operation (normalized to one cycle), but operates at a higher V_{DDopt}. Knowing the average workload and duty cycle of a circuit does impact the choice of optimum supply voltage. The operational frequency, and thus the data rate, implicitly changes with V_{DD} in these figures. A system in which these parameters vary widely would benefit from closed-loop tracking of the optimum point since Figure 4.15 and Figure 4.16 show a large variation in the minimum energy.

4.3.2 Temperature

The optimum point also depends on temperature. Figure 4.17 shows the effect of temperature on the components of energy. The numerical model shown in Figure 4.17 accounts for temperature dependence by decreasing the effective threshold voltage and decreasing mobility at higher temperatures: $\mu(T) = \mu(T_0)(\frac{T}{T_0})^{-M}$ and $V_T(T) = V_T(T_0) - KT$ [104]. The changes to the numerical model match well with simulation across most of the temperature range, but they slightly underestimate the leakage energy at high temperatures. The lower mobility dominates in strong inversion and leads to slower

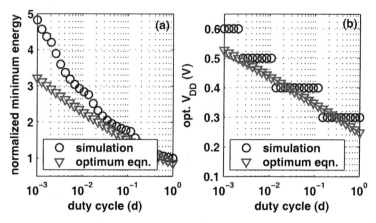

Fig. 4.16. Normalized energy (a) and V_{DDopt} (b) for FIR versus duty cycle, d. Simulation V_{DD} quantized to 100mV. (© 2004 IEEE)

circuits at high temperatures. In sub-threshold, the lower V_T dominates, and hot circuits grow faster exponentially. The lower V_T that accompanies a temperature increase also raises the leakage current exponentially. This effect appears in the figure at higher V_{DD} where I_{LEAK} dominates E_L. The lower V_T also causes the delay to decrease in sub-threshold, countering the increase in I_{LEAK} that is due to lower V_T. Thus, the total effect on E_L and E_T is not so pronounced at lower V_{DD}. Consequently, the total leakage energy does not change quickly near the minimum energy point with V_T. Equation (4.17) shows that V_{DDopt} is linear with temperature, and taking its derivative with respect to temperature gives:

$$\frac{\partial V_{DDopt}}{\partial T} = n\frac{k}{q}\left(2 - \text{lambertW}\left(\frac{-2C_{eff}}{W_{eff}KC_gL_{DP}}\exp(2)\right)\right) \quad (4.22)$$

For the FIR filter, this derivative is only ~ 0.85mV/degree Kelvin, and the plot confirms that the V_{DDopt} increases by about 75mV over the full temperature range.

4.3.3 Architecture

The models we have presented allow a quick assessment of the effect of architecture on minimum energy operation. Traditionally, architectural approaches such as parallelism or pipelining can reduce power for a given performance constraint by operating at a lower voltage [105]. For performance constraints that are met at voltages above the optimum, the same conclusion holds for sub-threshold operation. The model can also show how architecture will affect the minimum energy point when performance is not a constraint.

4.3 Minimum Energy Point Dependencies

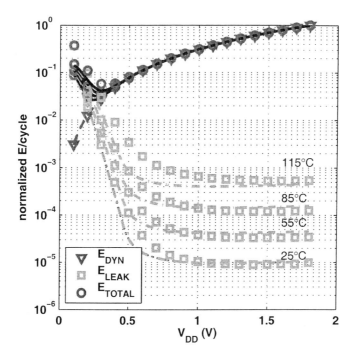

Fig. 4.17. Dependence of minimum energy point on temperature shown in simulation (markers) and by the numerical model (lines). Temperature varies from 25°C to 115°C. (© 2004 IEEE)

Figure 4.18 shows the numerical model (a) and simulation (b) of a pipelined implementation of the FIR filter. The model does not account for overhead capacitance, leakage, and delay in the pipeline registers, but it shows the general effect of ideal pipelining. As the number of stages increases, L_{DP} decreases and thus reduces leakage energy per operation (E_L). The total energy per cycle thus is reduced, and V_{DDopt} moves to the left. The simulation results in Figure 4.18 (b) show the same trend, however the overhead active energy makes deep pipelining more costly. The simulation also shows a decrease in active energy for shallow pipelines because of reduced glitching in the multipliers. Thus, shallow pipelining (2-4 stages) can reduce the total energy per operation for a system in a minimum energy scenario.

In contrast, parallelism cannot reduce the energy per operation, but it can increase the operating frequency at the minimum energy point. Once V_{DDopt} is known for a functional unit like the FIR filter, parallel copies of the filter will consume the least energy if they operate at the original minimum energy point. Discounting the overhead of muxing and demuxing indicates that ideal parallelism can increase the operating frequency at the minimum energy point

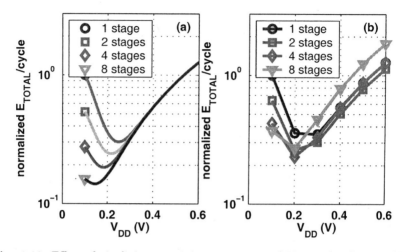

Fig. 4.18. Effect of pipelining on minimum energy and V_{DDopt} for the FIR filter. Ideal pipelines (a) and simulated (b). (© 2004 IEEE)

by increasing the number of stages. It cannot, however, decrease the energy per operation because the number of copies of the circuit required to maintain throughput goes up faster than V_{DD} decreases due to the exponential dependence of delay on V_{DD}. Clearly, the overhead of parallelism in a real system will increase active energy and change the minimum energy point.

5

EKV Model of the MOS Transistor

by Eric A. Vittoz and Christian C. Enz

5.1 Introduction and Definitions

The transistor model described in this chapter is the core of the EKV compact model that has been specially developed for low-voltage and/or low-current circuit design. It is a shortened description, focusing on weak and moderate inversion regimes that correspond to sub-threshold operation. More detailed derivations can be found in [45, 49, 106]. A very detailed presentation can be found in "Charge-Based MOS Transistor Modeling: The EKV Model for Low-Power and RF IC Design", by C. Enz and E. Vittoz, 2006, ©John Wiley & Sons Limited [52]. (Figures 1, 2, 4, 5, 6, 7, 10, 15, 16, 18 and 19 of this chapter are extracted from this book and reproduced with permission.)

The schematic cross-section of an N-channel MOS transistor with channel length L and channel width W is illustrated in Figure 5.1. In order to maintain

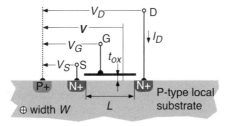

Fig. 5.1. Schematic cross-section of an N-channel MOS transistor, with definitions of voltages and current.

the intrinsic symmetry of the device, the source voltage V_S, gate voltage V_G and drain voltage V_D are all defined with respect to the local substrate. The drain current I_D is positive if it enters the drain. The *channel voltage* V (quasi-Fermi potential of electrons in the channel) changes monotonically, from $V = V_S$ at the source end of the channel to $V = V_D$ at the drain end of the channel.

Another important voltage is the thermodynamic voltage

$$U_T = kT/q, \tag{5.1}$$

where k is the Boltzmann constant and q the elementary charge. Proportional to the absolute temperature T, it is a measure of the thermal energy of electrons. Its value is 25.8 mV at 300 K (or 27 °C) and it appears ubiquitously in MOS modeling equations.

The doping concentration of the substrate is assumed to have a constant value N_b in the channel, and the gate oxide thickness t_{ox} corresponds to a capacitance $C_{ox} = \epsilon_{ox}/t_{ox}$ per unit area.

Figure 5.2 shows the symbols that can be used in order to preserve the symmetry of the device. It also shows how the definition of positive voltages and current can be inverted so that the model developed for the N-channel transistor can be applied without any change to the P-channel device.

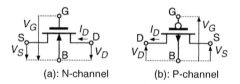

(a): N-channel   (b): P-channel

Fig. 5.2. Symbols for N-channel and P-channel devices with the respective definitions of positive voltages and current.

5.2 Density of Mobile Charge

5.2.1 Threshold Function

When a positive voltage is applied to the gate of the N-channel transistor, the holes in the channel are repelled away from the surface, thereby creating a depleted layer underneath the silicon surface and increasing its potential Ψ_s. This depleted layer is due to the remaining fixed ionized impurity atoms and is characterized by a negative charge density Q_b (per unit area of channel) given by

$$Q_b = -\Gamma_b C_{ox} \sqrt{\Psi_s}, \tag{5.2}$$

where

$$\Gamma_b = \frac{\sqrt{2qN_b\epsilon_{si}}}{C_{ox}} \tag{5.3}$$

is the substrate modulation factor. This fixed charge Q_b is useless since it cannot move to create a current. But the positive surface potential also attracts electrons to the surface, producing a mobile *inverted charge* of local density Q_i that can carry a current. The thickness of this inversion charge layer is very small, therefore the voltage drop across it can be neglected (charge sheet approximation).

Using Gauss' law, the total charge density underneath the silicon surface is given by

$$Q_{si} = Q_b + Q_i = -C_{ox}(V_G - V_{FB} - \Psi_s), \tag{5.4}$$

where V_{FB} is the flat-band voltage that includes the difference Φ_{ms} of extraction potentials between the gate and channel material and the effect of the fixed charge of density Q_{fc} possibly trapped in the oxide and at its interface with the silicon:

$$V_{FB} = \Phi_{ms} - Q_{fc}/C_{ox}. \tag{5.5}$$

Combining (5.4) with (5.2) yields the density of inverted charge

$$Q_i = -C_{ox}(V_G - V_{FB} - \Psi_s - \Gamma_b\sqrt{\Psi_s}) = -C_{ox}(V_G - V_{TB}), \tag{5.6}$$

where

$$V_{TB} = V_{FB} + \Psi_s + \Gamma_b\sqrt{\Psi_s} \tag{5.7}$$

is a *threshold function*. This function of the surface potential Ψ_s depends on the process through parameters Γ_b and V_{FB}. It is represented in Figure 5.3(a) for a particular value of Γ_b. The figure also shows the value of $-Q_i/C_{ox}$ according to (5.6) and that of $-Q_b/C_{ox}$ for a particular value of the gate voltage V_G. This function is nonlinear due to the contribution of Q_b, and its slope $n > 1$

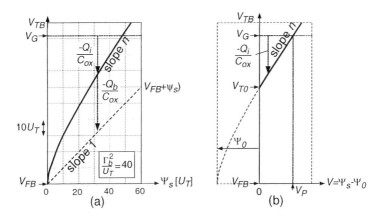

Fig. 5.3. Threshold function and inverted charge density: (a) as a function of the surface potential; (b) approximation in strong inversion.

is obtained by differentiation of (5.7):

$$n = \frac{dV_{TB}}{d\Psi_s} = 1 + \frac{\Gamma_b}{2\sqrt{\Psi_s}}. \qquad (5.8)$$

Inspection of Figure 5.3(a) shows that, for a fixed value of the gate voltage V_G,

$$\frac{dQ_i/C_{ox}}{d\Psi_s} = n. \qquad (5.9)$$

Now, the local value of inverted charge density Q_i can only be obtained from Figure 5.3(a) if the local value of Ψ_s is known. We shall first consider the case of strong inversion.

5.2.2 Approximation in Strong Inversion

It can be shown [107, 52] that the local inverted charge increases exponentially with $\Psi_s - V$ according to

$$Q_i \propto \exp \frac{\Psi_s - 2\Phi_F - V}{U_T}. \qquad (5.10)$$

where $U_T = kT/q$ is the thermodynamic voltage, and Φ_F is the Fermi potential of the substrate, which depends on its doping concentration and on the intrinsic carrier concentration of silicon n_i according to

$$\Phi_F = U_T \ln \frac{N_b}{n_i}. \qquad (5.11)$$

Hence, as soon as Q_i starts to dominate in strong inversion, the surface potential Ψ_s only increases very slowly since the total charge Q_{si} is limited by the limited field in the oxide. For this reason, the surface potential can be assumed to be independent of V_G and to be given by [52]

$$\Psi_s = V + \Psi_0, \qquad (5.12)$$

where

$$\Psi_0 = 2\Phi_F + \text{a few } U_T. \qquad (5.13)$$

In this approximation, the threshold function $V_{TB}(V)$ is therefore identical to $V_{TB}(\Psi_s)$, but its vertical axis ($V = 0$) is shifted by Ψ_0, as illustrated in Figure 5.3(b). For $V = 0$, V_{TB} takes the particular value V_{T0} called the *equilibrium threshold voltage*, or for short *threshold voltage*. Its expression is obtained by replacing Ψ_s by Ψ_0 in (5.7):

$$V_{T0} = V_{FB} + \Psi_0 + \Gamma_b \sqrt{\Psi_0}. \qquad (5.14)$$

This *bias-independent* device parameter corresponds to the threshold voltage V_T for $V_S = 0$ used in other models.

As shown by Figure 5.3(b), the slope n for $V > 0$ can be considered constant and will be called the *slope factor*. Now in this approximation, for a particular value of the gate voltage V_G, $Q_i = 0$ for a particular value V_P of V called the *pinch-off voltage*. Inspection of the figure shows that V_P is related to V_G by

$$V_P = \frac{V_G - V_{T0}}{n}, \tag{5.15}$$

and that Q_i can be expressed as

$$-Q_i/C_{ox} = n(V_P - V), \tag{5.16}$$

where the slope factor n is given by (5.8). It is usually convenient to evaluate it at $\Psi_s = \Psi_0 + V_P$, giving

$$n = 1 + \frac{\Gamma_b}{2\sqrt{\Psi_0 + V_P}}. \tag{5.17}$$

5.2.3 General Case

By differentiation of (5.10) we obtain

$$U_T \frac{dQ_i}{Q_i} = d\Psi_s - dV. \tag{5.18}$$

Now, introducing (5.9) in (5.18) to eliminate $d\Psi_s$ results in

$$\frac{dV}{U_T} = \frac{dQ_i}{nU_T C_{ox}} - \frac{dQ_i}{Q_i}. \tag{5.19}$$

In should be pointed-out that the assumption of constant n amounts to a *linearization* of the charge-potential relationship.

Further calculations can be simplified by normalizing voltage and charge according to

$$v = V/U_T \quad \text{and} \quad q_i = Q_i/Q_{spec}, \tag{5.20}$$

where

$$Q_{spec} = -2nC_{ox}U_T. \tag{5.21}$$

Equation (5.19) then becomes

$$-dv = 2dq_i + dq_i/q_i. \tag{5.22}$$

Integrating both sides of this equation yields

$$\text{constant} - v = 2q_i + \ln q_i. \tag{5.23}$$

Now, in strong inversion, $\ln q_i \ll 2q_i$. The comparison with (5.16) after de-normalization shows that the constant in (5.23) is equal to $v_p = V_P/U_T$, hence

$$v_p - v = 2q_i + \ln q_i, \tag{5.24}$$

which is the general relationship between voltages and mobile inverted charge density. This relation is plotted in Figure 5.4 but, in the general case, it cannot be inverted to obtain the charge from the voltages.

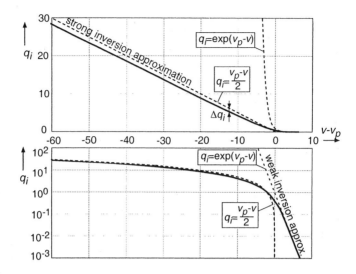

Fig. 5.4. Normalized inverted charge vs. normalized channel voltage.

5.2.4 Approximation in Weak Inversion

In weak inversion, $2q_i \ll |\ln q_i|$, therefore (5.24) can be approximated by

$$q_i = \exp(v_p - v) \quad \text{or} \quad -Q_i/C_{ox} = 2nU_T \exp\frac{V_P - V}{U_T}, \qquad (5.25)$$

where the V_P is the pinch-off voltage defined by (5.15) in the strong inversion approximation.

5.3 Drain Current and Modes of Operation

5.3.1 Charge-Current Relationship

The drain current I_D is the sum of conduction and diffusion currents given by [108]

$$I_D = \mu W \left(\underbrace{-Q_i \frac{d\Psi_s}{dx}}_{\text{conduction}} + \underbrace{U_T \frac{dQ_i}{dx}}_{\text{diffusion}} \right), \qquad (5.26)$$

where μ is the carrier mobility, and x is the position along the channel starting from the source side. Now, by introducing (5.18), this relation becomes

$$I_D = \mu W (-Q_i) \frac{dV}{dx}. \qquad (5.27)$$

5.3 Drain Current and Modes of Operation

Assuming constant mobility, integrating this equation along the channel gives

$$I_D = \beta \int_{V_S}^{V_D} \frac{-Q_i}{C_{ox}} dV, \quad (5.28)$$

where

$$\beta = \mu C_{ox} W/L \quad (5.29)$$

is the transfer parameter of the transistor. Hence, the drain current is proportional to the integral from $V = V_S$ to $V = V_D$ of the $Q_i(V)$ function obtained in Section 5.2, as represented in Figure 5.5(a).

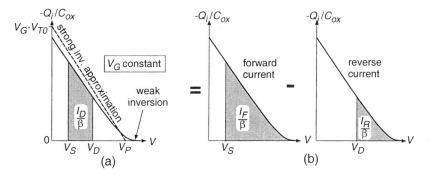

Fig. 5.5. (a) Drain current; (b) decomposition in forward and reverse components.

5.3.2 Forward and Reverse Components

Since Q_i tends to zero for V tending to infinity, the integral (5.28) can be rewritten as

$$I_D = \underbrace{\beta \int_{V_S}^{\infty} \frac{-Q_i}{C_{ox}} dV}_{\text{forward current } I_F} - \underbrace{\beta \int_{V_D}^{\infty} \frac{-Q_i}{C_{ox}} dV}_{\text{reverse current } I_R} = I_F - I_R. \quad (5.30)$$

Hence, as illustrated by Figure 5.5(b), the drain current can be expressed as the difference between a *forward current* I_F and a *reverse current* I_R. I_F depends on V_G and V_S, but *not on* V_D, whereas I_R depends on V_G and V_D, but *not on* V_S. Furthermore, according to (5.30), $I_F(V_S) \equiv I_R(V_D)$: I_F and I_R are indeed two values of the same function of V. Thus, the drain current is the *superposition* of *independent* and *symmetrical* effects of the source and drain voltages [109].

5.3.3 General Current Expression

By introducing the normalized variable defined by (5.20), the forward or reverse components defined by (5.30) can be expressed in normalized form as

$$i_{f,r} = \frac{I_{F,R}}{I_{spec}} = \int_{v_{s,d}}^{\infty} q_i dv, \qquad (5.31)$$

where $v_{s,d}$ is the source or drain voltage normalized to U_T, and

$$I_{spec} = 2n\mu C_{ox}\frac{W}{L}U_T^2 = 2n\beta U_T^2 \qquad (5.32)$$

is the *specific current* of the transistor.

Introducing (5.22) into (5.31) yields

$$i_{f,r} = \int_0^{q_{s,d}} (2q_i + 1)dq_i = q_{s,d}^2 + q_{s,d}, \qquad (5.33)$$

where $q_{s,d}$ is the value of the normalized charge density q_i at the source end or at the drain end of the channel. Solving this equation for the charge gives

$$q_{s,d} = \frac{\sqrt{1 + 4i_{f,r}} - 1}{2}. \qquad (5.34)$$

Using the voltage-charge relationship (5.24), we obtain finally

$$v_p - v_{s,d} = \sqrt{1 + 4i_{f,r}} + \ln\left(\sqrt{1 + 4i_{f,r}} - 1\right) - (1 + \ln 2) \qquad (5.35)$$

This general expression of the current-voltage relationship is plotted in Figure 5.6 (curve a), by calculating the voltages from the current. This figure also shows the approximation in strong inversion (curve b for $i_{f,r} \gg 1$) and that in weak inversion (curve c for $i_{f,r} \ll 1$). Remembering that $i_d = i_f - i_r$ and $v_p = (v_g - v_{t0})/n$, (5.35) models the static transistor characteristics from weak to strong inversion with only 3 model parameters (besides U_T used to normalize all voltages): the threshold voltage V_{T0}, the slope factor n and the specific current I_{spec} (used to normalized the currents) that includes the transfer parameter β according to (5.32).

Now, since (5.35) cannot be inverted to calculate the current from the voltages, it can be approximated by [47, 48, 110]

$$i_{f,r} = \ln^2\left(1 + \exp\frac{v_p - v_{s,d}}{2}\right), \qquad (5.36)$$

which is also plotted in Figure 5.6 (curve d).

5.3 Drain Current and Modes of Operation   57

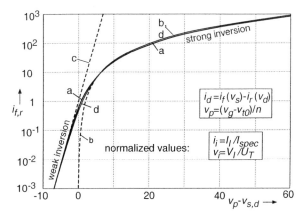

Fig. 5.6. Normalized forward or reverse current ; (a) from charge model (5.33); (b) strong inversion approximation; (c) weak inversion approximation; d) from interpolation formula (5.36).

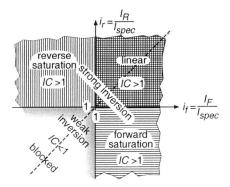

Fig. 5.7. Modes of operation of a MOS transistor.

5.3.4 Modes of Operation and Inversion Coefficient

The various possible modes of operation of a transistor depend on the values of I_F and I_R. They can be described in the (i_f, i_r) plane represented in Figure 5.7. Although weak and strong inversion are separated by a regime of moderate inversion (c.f. Figure 5.6), we shall simplify the discussion by assuming that $i_{f,r} = 1$ ($I_{F,R} = I_{spec}$) represents the limit between weak and strong inversion for each of the two components.

If $i_f > 1$ and $i_r > 1$, then both components are in strong inversion and the whole channel is strongly inverted. The transistor is said to be in *linear* mode.

If $i_f > 1$ but $i_r < 1$, the reverse component is negligible and the current does not increase anymore with the drain voltage: the transistor is still in strong inversion, but in *forward saturation*. If $i_r > 1$ but $i_f < 1$, the forward component is negligible and the current does not increase anymore with the source voltage: the transistor is still in strong inversion, but in *reverse saturation* ($i_d < 0$).

If $i_f < 1$ and $i_r < 1$, then both components are in weak inversion, and the *whole channel is only weakly inverted*. The transistor is said to operate in *weak inversion*.

The global level of inversion of the transistor can be characterized by its *inversion coefficient IC* defined by

$$IC = \max(i_f, i_r) \tag{5.37}$$

The transistor operates in weak inversion for $IC \ll 1$, in strong inversion for $IC \gg 1$, and in moderate inversion for $IC \cong 1$.

5.3.5 Output Characteristics and Saturation Voltage

In forward mode, $IC = i_f$. If the drain voltage is increased, the reverse current is progressively decreased until it becomes negligible. Therefore, the drain current $i_d = i_f - i_r$ increases until it saturates at the value $i_{dsat} = i_f$. The drain to source voltage $v_{ds} = v_d - v_s$ can be obtained by calculating $(v_p - v_s) - (v_p - v_d)$ from (5.35), with $i_r = IC(1 - i_d/i_f)$. This yields

$$v_{ds} = \sqrt{1+4IC} - \sqrt{1+4IC(1-i_d/i_f)} + \ln\frac{\sqrt{1+4IC}-1}{\sqrt{1+4IC(1-i_d/i_f)}-1}. \tag{5.38}$$

This expression is plotted in Figure 5.8 for several values of the inversion coefficient IC ranging from weak inversion to strong inversion.

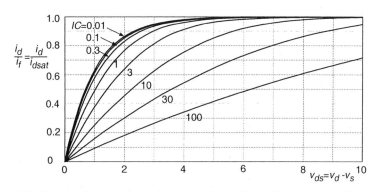

Fig. 5.8. Output characteristics for increasing values of inversion coefficient IC.

As can be seen, weak inversion ($IC \ll 1$) provides the minimum possible saturation voltage, since the drain current saturates for $V_{DS} \approx 5U_T$. This is why it is intrinsically associated with very *low voltage circuit* design. The

5.4 Small-Signal Model

saturation voltage starts increasing in moderate inversion and tends to $V_P - V_S = 2U_T\sqrt{IC}$ in very strong inversion.

5.3.6 Weak Inversion Approximation

The components of the drain current in weak inversion can be obtained by integrating the charge density given by (5.25) according to (5.31). This yields

$$i_{f,r} = \exp(v_p - v_{s,d}) \quad \text{or} \quad I_{F,R} = I_{spec} \exp \frac{V_P - V_{S,D}}{U_T} \quad (5.39)$$

which is plotted as curve c in Figure 5.6. It is only exact for $IC = \max(i_f, i_r) \ll 1$. Then by introducing the expression (5.15) of V_P:

$$I_D = I_{spec} \exp \frac{V_G - V_{T0}}{nU_T} \left(\exp \frac{-V_S}{U_T} - \exp \frac{-V_D}{U_T} \right). \quad (5.40)$$

The first term in the parenthesis belongs to the forward mode, the second term belongs to the reverse mode. The latter becomes negligible as soon as V_D exceeds V_S by a few U_T, as seen in Figure 5.8. The slope factor n in the common exponential term represents the effect of the capacitive divider formed by the oxide capacitance C_{ox} and the depletion capacitance C_d.

This equation can be rewritten by lumping the dependencies on I_{spec} and V_{T0}

$$I_D = I_{D0} \exp \frac{V_G}{nU_T} \left(\exp \frac{-V_S}{U_T} - \exp \frac{-V_D}{U_T} \right), \quad (5.41)$$

where

$$I_{D0} = I_{spec} \exp \frac{-V_{T0}}{nU_T} \quad (5.42)$$

is the residual drain current in saturation for $V_G = V_S = 0$, or channel "leakage" current of CMOS digital circuits. This current increases exponentially when the threshold voltage is reduced.

5.4 Small-Signal Model

5.4.1 Transconductances

The DC small-signal equivalent circuit of the 4-terminal transistor is shown in Figure 5.9. Small variations of V_S, V_D and V_G produce small variations of the drain current proportional to the respective transconductances G_{ms} (source transconductance), G_{md} (drain transconductance) and G_m (gate transconductance).

The drain current I_D depends on V_S through I_F only and on V_D through I_R only. Hence, the source and drain transconductances are obtained by differentiating (5.35) and then inverting and de-normalizing the result. This gives

5 EKV Model of the MOS Transistor

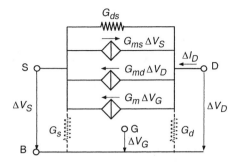

Fig. 5.9. DC small-signal equivalent circuit.

$$G_{ms,d} = \frac{I_{spec}}{2U_T}\left(\sqrt{1 + 4I_{F,R}/I_{spec}} - 1\right). \tag{5.43}$$

At small and large current, it tends to

$$G_{ms,d} = \frac{I_{F,R}}{U_T} \text{ (weak inversion)}, \tag{5.44}$$

$$G_{ms,d} = \frac{\sqrt{I_{F,R} \cdot I_{spec}}}{U_T} = \sqrt{2n\beta I_{F,R}} \text{ (strong inversion)}. \tag{5.45}$$

Now, since $I_{F,R}$ is a function of $V_P - V_{S,D}$, the gate transconductance is given by

$$G_m = \frac{\partial I_D}{\partial V_G} = \frac{\partial(I_F - I_R)}{\partial V_P} \cdot \frac{\partial V_P}{\partial V_G} = -\left(\frac{\partial I_F}{\partial V_S} - \frac{\partial I_R}{\partial V_D}\right)\frac{1}{n} = \frac{G_{ms} - G_{md}}{n}. \tag{5.46}$$

In (forward) saturation $I_R \ll I_F$, hence $G_{md} \ll G_{ms}$ and $G_m = G_{ms}/n$.

Since the transconductance always increases with the current, it is interesting to express the transconductance-to-current ratio. Dividing (5.43) by $I_{F,R}$ gives

$$\frac{G_{ms,d}}{I_{F,R}} = \frac{1}{U_T} \cdot \frac{2}{\sqrt{1 + 4I_{F,R}/I_{spec}} + 1}. \tag{5.47}$$

This result is plotted in Figure 5.10 for G_{ms} (curve a). In weak inversion G_{ms}/I_F reaches the maximum possible value $1/U_T$. It is reduced by about 40 % at $IC = 1$ and tends to $1/(U_T\sqrt{IC})$ in strong inversion.

Curve b in the same figure has been obtained by differentiating the approximation (5.36) of the drain current, resulting in

$$\frac{G_{ms,d}}{I_{F,R}} = \frac{1}{U_T} \cdot \frac{1 - \exp\left(-\sqrt{I_{F,R}/I_{spec}}\right)}{\sqrt{I_{F,R}/I_{spec}}}. \tag{5.48}$$

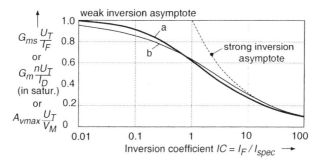

Fig. 5.10. Variation of the transconductance-to-current ratio with the inversion coefficient; (a) Charge-based model (5.47); (b) Approximation (5.48) obtained from (5.36).

5.4.2 Residual Conductance in Saturation and Maximum Voltage Gain

The conductance G_{ds} included in the small-signal equivalent circuit of Figure 5.9 represents the residual output conductance in saturation that is due to channel length modulation by the drain voltage (in forward mode). This conductance limits the maximum possible voltage gain in common-gate configuration to

$$A_{vmax} = G_{ms}/G_{ds}, \quad (5.49)$$

whereas this maximum value is reduced by n in common-source configuration.

In forward mode, G_{ds} is approximately proportional to the saturation current I_F according to

$$G_{ds} = I_F/V_M, \quad (5.50)$$

where V_M is the *channel length modulation voltage*, proportional to the channel length. Therefore, Figure 5.10 also represents the variation of A_{vmax} with the inversion coefficient, with a maximum value V_M/U_T in weak inversion.

Conductances G_d and G_s included in the equivalent circuit of Figure 5.9 are the differential conductances of the reverse biased drain and source junctions. Their value is small but independent of the drain current. They may therefore become larger than G_{ds} at very small current, thereby dominating the output conductance and limiting the voltage gain.

5.4.3 Small-Signal AC Model

For frequencies below $\mu U_T/L^2$, the dynamic behavior of the transistor can be modeled by means of lumped capacitors added to the small-signal model.

Intrinsic capacitors are due to the charge stored in the channel. Each of them is a bias-dependent fraction of the total gate oxide capacitor WLC_{ox}.

Extrinsic capacitors are those of the source and drain junctions, and the gate overlap capacitors to source and drain diffusions. Their values are essentially independent of the drain current.

In *weak inversion*, most of the intrinsic capacitances are negligible if the channel is not very long. The only non-negligible value is the gate-to-substrate capacitance produced by the series connection of C_{ox} and the depletion capacitance C_d. It is given by

$$C_{GB} = \frac{n-1}{n} \cdot WLC_{ox}, \qquad (5.51)$$

and is smaller that the total gate oxide capacitance.

5.5 Transistor Operated As a Pseudo-Resistor

The appellations source and drain for the two diffused regions forming a transistor are purely functional (the source and drain functions being inverted if the voltage is inverted) and can normally not be identified in the structure itself. Let us therefore name them simply A and B to emphasize the symmetry of the device, as shown in Figure 5.11.

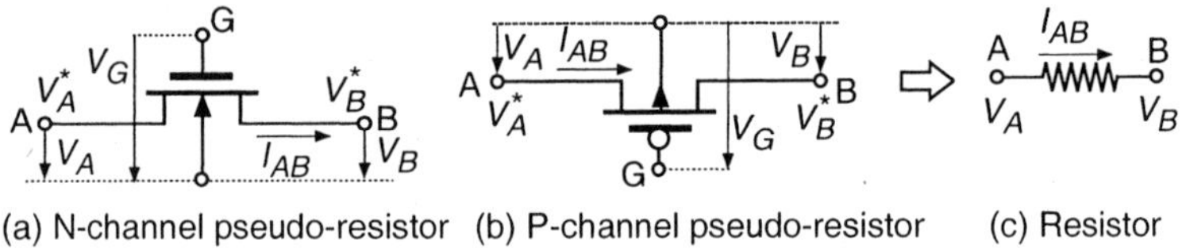

(a) N-channel pseudo-resistor   (b) P-channel pseudo-resistor   (c) Resistor

Fig. 5.11. Pseudo-resistors: (a) N-channel; (b) P-channel; (c) prototype resistor.

The expression (5.40) of the drain current in weak inversion then becomes

$$I_{AB} = \pm I_{spec} \exp \frac{V_G - V_{T0}}{nU_T} \left(\exp \frac{-V_A}{U_T} - \exp \frac{-V_B}{U_T} \right), \qquad (5.52)$$

which is a linear relationship between the current and the exponential function of the voltages. The $-(+)$ sign applies to the N-channel (P-channel) transistor. Let us define *pseudo-voltages* [111, 112, 109] corresponding to V_A and V_B as

$$V_{A,B}^\star = \pm V_0 \exp \frac{-V_{A,B}}{U_T}, \qquad (5.53)$$

and a *pseudo-conductance*

$$G^\star = \frac{1}{R^\star} = \frac{I_{spec}}{V_0} \exp \frac{V_G - V_{T0}}{nU_T}, \qquad (5.54)$$

where V_0 is any positive voltage. Then (5.52) becomes

$$I_{AB} = G^\star(V_A^\star - V_B^\star) = (V_A^\star - V_B^\star)/R^\star, \tag{5.55}$$

which is a linear *pseudo-Ohm's law*. Hence, by similarity with a network of linear resistors, any network obtained by interconnecting transistors by their source and drain is *linear for currents* (and for pseudo-voltages, but *not* for voltages). In other words, at each node of such a network, the current splits linearly in the various branches [113].

Thus, any prototype made of real linear resistors may be converted to a pseudo-resistor network made of transistors only, by replacing each resistor by the source-drain port of a transistor. Moreover, each pseudo-resistance is controllable by the gate voltage of the transistor, according to (5.54). It must be noticed that the general principle is also valid in moderate and strong inversion [113, 111, 112, 109], but then the gate voltages must be identical for all transistors. Therefore, the possibility to control $R^\star$ by the gate voltage only exists in weak inversion (I_F and $I_R \ll I_{spec}$).

It must be pointed out that no voltage should be applied or measured in such current-mode circuits. Therefore, the value of V_0 in (5.53) and (5.54) is irrelevant.

If one side of the transistor is saturated, the corresponding pseudo-voltage is zero. It is a *pseudo-ground* $0^\star$ that corresponds to a ground ($V = 0$) in the resistor prototype. Notice that, according to (5.53) pseudo-voltages for a N-channel (P-channel) are always negative (positive).

When used in weak inversion, the concept of pseudo-resistors is only degraded by channel-length modulation and short-channel effects. Examples of application will be given in Section 8.5.

5.6 Noise

5.6.1 Noise model

Noise is introduced in the model by adding two noise sources to the noiseless transistor as shown in Figure 5.12. The channel noise is modeled by a noise

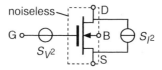

Fig. 5.12. Modeling the noise by two noise sources of power spectral densities S_{I^2} and S_{V^2}.

current source of power spectral density S_{I^2} (dimension A^2/Hz). The gate interface noise is best modeled by a voltage noise source of power spectral density S_{V^2} (dimension V^2/Hz).

5.6.2 Channel Noise

Channel noise is the most fundamental noise. In weak inversion, it is a shot noise [114] associated with the barrier that controls the amount of carrier diffusing in the channel. It can be shown that independent noise sources are associated with the source and drain barriers and that their respective spectral densities are $2qI_F$ and $2qI_R$ [52]. Hence, in weak inversion

$$S_{I^2} = 2q(I_F + I_R) = 2qI_F \left(1 + \exp\frac{-V_{DS}}{U_T}\right), \tag{5.56}$$

which is plotted in Figure 5.13.

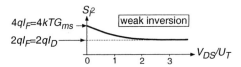

Fig. 5.13. Power spectral density of channel noise in weak inversion.

For $V_{DS} \gg U_T$ (saturated weak inversion), $I_D = I_F$ and

$$S_{I^2} = 2qI_D, \tag{5.57}$$

whereas for $V_{DS} = 0$, $I_D = I_F - I_R = 0$ and

$$S_{I^2} = 4qI_F = 4kT \cdot G_{ms}. \tag{5.58}$$

The spectral density is doubled when V_{DS} is reduced from saturation to zero. For $V_{DS} = 0$, it is equal to that of the channel conductance $G_{ms} = G_{md}$. The latter result is also true for strong inversion, but S_{I^2} is then reduced to 2/3 (instead of 1/2) when saturation is reached, for $V_{DS} \geq V_P$. Hence, by introducing expression (5.45) of the transconductance in strong inversion, in saturation ($I_D = I_F$)

$$S_{I^2} = 4kT \cdot \frac{2}{3} \cdot \frac{\sqrt{I_F I_{spec}}}{U_T} = 2qI_D \cdot \frac{4}{3\sqrt{IC}}, \tag{5.59}$$

which is decreased by increasing the inversion coefficient IC.

5.6.3 Interface Noise

The interface noise is due to a combination of carrier number fluctuation due to interface traps and surface mobility fluctuation. Its spectral density can be globally modeled by [110]

$$S_{V^2} = \frac{4kT\rho}{f \cdot WL} \tag{5.60}$$

in order to express its dependency on the inverse of both the frequency f and the channel area WL. The parameter ρ depends on the process and on the gate oxide capacitance ($\rho \propto C_{ox}^{-\alpha}$, with $1 < \alpha < 2$). Its value is somewhat dependent on the inversion coefficient, with a flat minimum around $IC = 1$ [52].

5.6.4 Total Noise

In saturation, the spectral density of the total *output* current noise for constant gate voltage is

$$S_{I_D^2} = S_{I^2} + G_m^2 S_{V^2}. \tag{5.61}$$

At a given current I_D, $S_{I_D^2}$ is *maximum* in weak inversion since both S_{I^2} and G_m are maximum. But the spectral density of the total *input* referred noise voltage for constant drain current is

$$S_{V_G^2} = S_{V^2} + \frac{S_{I^2}}{G_m^2} = S_{V^2} + \frac{4kT\gamma_n}{G_m} = 4kTR_n, \tag{5.62}$$

where γ_n is the thermal noise excess factor, equal to $n/2$ in weak inversion and to $2n/3$ in strong inversion, and R_n is the input referred *equivalent noise resistance*. At a given current I_D, $S_{V_G^2}$ is *minimum* in weak inversion since G_m is maximum. The equivalent noise resistance is obtained by introducing (5.60) in (5.62):

$$R_n = \frac{\gamma_n}{G_m} + \frac{\rho}{f \cdot WL}. \tag{5.63}$$

5.7 Temperature effects

The dependence on temperature of the transistor characteristics can be modeled through that of the main parameters V_{T0}, n and β (and therefore also $I_{spec} = 2n\beta U_T^2$).

Essentially V_{T0} and n variation is due to Φ_F (Equation 5.11). It includes the direct effect of U_T but also the variation of the intrinsic carrier according to [107]

$$n_i \propto T^{3/2} \exp \frac{-V_{gap}}{2U_T} \tag{5.64}$$

where T is the absolute temperature and V_{gap} the voltage corresponding to the band gap of silicon. By linearizing $V_{gap}(T)$ at some ambient temperature T_0, (5.64) becomes

$$n_i \propto T^{3/2} \exp \frac{-(V_{G0} - aT)}{2U_T} \propto T^{3/2} \exp \frac{-V_{G0}}{2U_T} \quad (5.65)$$

where $-a$ is the slope of the tangent to $V_{gap}(T)$ at $T = T_0$ and V_{G0} is the value obtained by extrapolating this tangent to $T = 0$, called the *extrapolated band gap voltage*. Its value of about 1.2 V is only slightly dependent on T_0 due to the very small curvature of $V_{gap}(T)$. By neglecting the $T^{3/2}$ dependence with respect to the exponential dependence this gives finally

$$n_i = n_{i\infty} \cdot \exp \frac{-V_{G0}}{2U_T} \quad (5.66)$$

where $n_{i\infty}$ is a constant (extrapolated value of n_i at $T = T_0$). Introducing this expression in (5.11) gives

$$\Phi_F = \frac{V_{G0}}{2} - U_T \ln \frac{n_{i\infty}}{N_b} \quad \text{and} \quad \frac{d\Phi_F}{dT} = -\frac{1}{T}\left(\frac{V_{G0}}{2} - \Phi_F\right). \quad (5.67)$$

Now, if the gate is highly doped, the flat-band voltage V_{FB} defined by (5.5) includes $-\Phi_F$ (inside Φ_{ms}), which makes it dependent on temperature. If we neglect the small difference in (5.13) between Ψ_0 and $2\Phi_F$, the temperature dependence of V_{T0} obtained from (5.14) is

$$\frac{dV_{T0}}{dT} = \left(1 + \frac{\Gamma_b}{\sqrt{\Psi_0}}\right)\frac{d\Phi_F}{dT} = (2n_0 - 1)\frac{d\Phi_F}{dT} \quad (5.68)$$

where n_0 is the slope factor n evaluated at $V = 0$ (or $\Psi_s = \Psi_0$). Then, by combination with (5.67):

$$\frac{dV_{T0}}{dT} = \frac{n_0 - 0.5}{T}(2\Phi_F - V_{G0}) < 0. \quad (5.69)$$

The temperature coefficient of V_{T0} is always negative, with practical values ranging from -2.5 to -1 mV/°K.

If n is evaluated at $V = V_P$ ($\Psi_s = \Psi_0 + V_P \cong 2\Phi_F + V_P$), its temperature coefficient is given by

$$\frac{dn}{dT} = \frac{d}{d\Phi_F}\left(1 + \frac{\Gamma_b}{2\sqrt{2\Phi_F + V_P}}\right)\frac{d\Phi_F}{dT} = \frac{n-1}{2T} \cdot \frac{V_{G0} - 2\Phi_F}{2\Phi_F + V_P} > 0. \quad (5.70)$$

The temperature coefficient of n is always positive, with practical values below 10^{-3}/°K ($< 0.1\%$/°K). It can therefore be neglected for $V_S = 0$. But for $V_S > 0$ it affects the effective threshold $V_{T0} + nV_S$, the temperature coefficient of which is improved by the opposite signs in (5.69) and (5.70).

The variation of β with temperature is due to that of the mobility μ that can be approximated, in the range of ambient temperatures, by

$$\mu \propto T^{-\alpha}, \tag{5.71}$$

where $1.5 < \alpha < 3$ depends on the doping concentration N_b. Hence, from the expression (5.29) of β:

$$\frac{\mathrm{d}\beta/\beta}{\mathrm{d}T} = -\frac{\alpha}{T} < 0. \tag{5.72}$$

The temperature coefficient of β is always negative, with practical values ranging from -0.5 to $-1\%/^\circ\mathrm{K}$.

It is worth noticing that for $\alpha = 2$, the specific current I_{spec} defined by (5.32) is independent of T since the variation of μ compensates that of U_T^2.

5.8 Non-ideal effects

5.8.1 Mismatch

Two or more transistors of identical structures implemented on the same chip do not have exactly the same characteristics. This is due to small differences in their dimensions and/or to variations of physical parameters. These differences are reflected in the model parameters essentially as differences of V_{T0}, n and β.

Assuming no temperature difference, the physical parameters possibly responsible for the mismatch of these three model parameters are identified in Table 5.1 on the basis of their respective equations.

Table 5.1. Physical parameters affecting the mismatch of the model parameters.

Equ.		Q_{fc}	N_b	C_{ox}	μ	W	L
(5.14)	ΔV_{T0}	•	•	•			
(5.8)	Δn		•	•			
(5.29)	$\Delta \beta$			•	•	•	•

Systematic differences, due for example to gradients of physical parameters across the chip, can be eliminated by means of adequate layout techniques [115]. However, some random mismatch remains due to random fluctuations of the parameters.

It can be shown [116] that the standard deviation of the difference ΔP of average values of a parameter P across two separate regions of area WL is given by

$$\sigma(\Delta P) = \frac{A_P}{\sqrt{WL}}, \tag{5.73}$$

where A_P is the area proportionality constant for parameter P.

5 EKV Model of the MOS Transistor

The picture is somewhat different for the dimensions W and L. Indeed, it can be shown [116, 52] that the standard deviation of their ratio W/L, that affects $\Delta\beta$, is given by

$$\sigma(\Delta(W/L)) = \frac{A_{WL}}{\sqrt{WL}} \cdot \sqrt{\frac{1}{W} + \frac{1}{L}}. \qquad (5.74)$$

Hence, this contribution to β-mismatch is proportional to $(WL)^{-3/2}$ for constant W/L. It can therefore be made negligible with respect to those of ΔC_{ox} and $\Delta\mu$ by increasing W and L.

As shown by Table 5.1, the mismatch of all three parameters may be correlated through ΔC_{ox}, whereas ΔV_{T0} and Δn may be further (positively) correlated through ΔN_b.

Let us assume that ΔV_{T0} and $\Delta\beta$ are not correlated (ΔC_{ox} negligible), and that Δn can be neglected. If two saturated transistors are biased at the same source and gate voltages (as in a current mirror), the standard deviation of their relative difference of drain currents is given by

$$\frac{\sigma(\Delta I_D)}{I_D} = \sqrt{\sigma^2(\Delta\beta) + \left[\frac{G_m}{I_D} \cdot \sigma(\Delta V_{T0})\right]^2}, \qquad (5.75)$$

since a small ΔV_{T0} results in $\Delta I_D = -G_m \Delta V_{T0}$.

If, on the contrary, the two drain currents are imposed (or expected) to be equal, then the standard deviation of the difference of gate voltages required to compensate the mismatch of parameters (input offset voltage) is

$$\sigma(\Delta V_G) = \sqrt{\sigma^2(\Delta V_{T0}) + \left[\frac{I_D}{G_m} \cdot \frac{\sigma(\Delta\beta)}{\beta}\right]^2}. \qquad (5.76)$$

The different weightings of $\sigma(\Delta\beta)$ and $\sigma(\Delta V_{T0})$ depend on G_m/I_D, which in turn depends on the inversion coefficient according to (5.47) [115]. The result is represented in Figure 5.14 for particular values of mismatch.

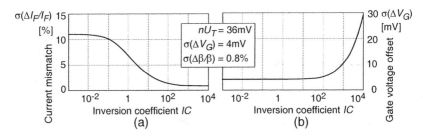

Fig. 5.14. Mismatch of (a) drain currents and (b) gate voltages for non-correlated $\sigma(\Delta V_{T0}) = 4mV$ and $\sigma(\Delta\beta)/\beta = 0.8\%$, and for $\sigma(\Delta n)$ negligible.

5.8 Non-ideal effects

As can be seen, the matching of currents is very bad in weak inversion (where it tends to $\sigma(\Delta I_D)/I_D = \sigma(\Delta V_{T0})/(nU_T)$). It is improved by increasing IC to reach $\sigma(\Delta\beta)/\beta$ in strong inversion.

On the contrary, the mismatch of gate voltages is limited to $\sigma(\Delta V_{T0})$ in weak inversion, but it is progressively degraded for IC increasing.

In practice, Δn is always negligible for $V_S = 0$. But for $V_S > 0$ it affects the effective threshold $V_{T0} + nV_S$, the matching of which is degraded by the positive correlation between $\sigma(\Delta V_{T0})$ and $\sigma(\Delta n)$ (through ΔC_{ox} and/or ΔN_b).

The minimum possible fluctuation in N_b is due to the limited number of impurities in the depletion volume WLt_d (where t_d is the thickness of the depleted layer). If this effect dominates, then $\sigma(\Delta V_{T0})$ and $\sigma(\Delta n)$ are fully correlated and can be calculated by assuming a Poisson's distribution of impurities, giving [52]

$$A_{VT0} = \sqrt{WL} \cdot \sigma(\Delta V_{T0}) = \frac{1}{C_{ox}} \sqrt[4]{q^3 \epsilon_{si} N_b \Phi_F}, \tag{5.77}$$

and

$$A_n = \sqrt{WL} \cdot \sigma(\Delta n) = \frac{A_{VT0}}{4\Phi_F} = \frac{1}{4C_{ox}} \sqrt[4]{\frac{q^3 \epsilon_{si} N_b}{\Phi_F^3}}. \tag{5.78}$$

Therefore

$$\sigma(V_{T0} + nV_S) = \sigma(V_{T0}) \left(1 + \frac{V_S}{4\Phi_F}\right). \tag{5.79}$$

As a consequence, the mismatch of gate voltages in weak inversion would be doubled for $V_S = 4\Phi_F = 1.2$ to 2 V.

5.8.2 Polysilicon Gate Depletion

In calculating the total charge density Q_{si} by (5.4), we have implicitly assumed a constant potential V_G throughout the thickness of the gate electrode, which would always be true if the gate material were metal. It is still true for a polysilicon gate, as long as the thickness of the layer of positive charge Q_g (negative for a P-channel transistor) is so small that the voltage drop across it is negligible. If the gate is P-type (N-type for a P-channel), the layer is still very thin since it is formed by majority carriers. But if the gate is N-type (P-type for a N-channel), Q_g is entirely produced by the depletion layer created at the lower face of the gate.

As long as the doping concentration N_g of the gate is much larger than N_b, the voltage drop across this depletion layer is negligible compared to the surface potential. Now, while scaling down process dimensions, N_b must be increased whereas N_g cannot be increased proportionally. The voltage drop in the gate is therefore no longer negligible and Q_i is no longer proportional to the difference between the gate voltage V_G and the threshold function $V_{TB}(\Psi_s)$.

Hence, the diagram of Figure 5.3 no longer applies, and the slope factor of $Q_i(\Psi_s)/C_{ox}$ in (5.9) is no longer the same as that of $V_G(V_P)$ in (5.15). The single slope factor n must be replaced by two distinct slope factors:

$$n_q = \frac{\mathrm{d}Q_i/C_{ox}}{\mathrm{d}\Psi_s} \quad \text{and} \quad n_v = \frac{\mathrm{d}V_G}{\mathrm{d}V_P}. \tag{5.80}$$

If the fixed charge density Q_{fc} is negligible, these slope factors can be related to n by [52]

$$n_q = n - \frac{1}{1 + (n-1)N_g/N_b} < n \tag{5.81}$$

and

$$n_v = n + N_b/N_g > n. \tag{5.82}$$

Moreover, the threshold voltage is increased to

$$V_{T0}^+ = V_{T0} + \frac{N_b}{N_g} \cdot \Psi_0. \tag{5.83}$$

The situation is more complicated if Q_{fc} is not negligible, since it can no longer be included in a constant flat-band voltage V_{FB} [52].

The pinch-off voltage is now obtained by replacing n by n_v and V_{T0} by V_{T0}^+ in (5.15). It is thus reduced, since $V_{T0}^+ > V_{T0}$ and $n_v > n$. All normalized equations developed previously are applicable, provided n is replaced by n_q in the definitions of Q_{spec} (5.21) and I_{spec} (5.32). Both of them are thus reduced (in absolute value) since $n_q < n$.

The source and drain transconductances for a given current are reduced in strong inversion (5.45) but not in weak inversion (5.44). The gate transconductance is always reduced, since n must be replaced by $n_v > n$ in (5.46).

5.8.3 Band Gap Widening

Since scaling down a process requires an increase of N_b, the electric field required at the surface to produce inversion is increased, requiring an increase of C_{ox}. As a consequence of this high field, the lowest allowed energy level for electrons in the conduction band is increased, which corresponds to a widening of the band gap.

It can be shown [52] that this band gap widening effect results in an increase of Ψ_0 with respect to its original value (5.13) by

$$\Delta\Psi_0 = A_{qm} \left[2q\epsilon_{si} N_b (\Psi_0 + V_P) \right]^{1/3}, \tag{5.84}$$

with $A_{qm} = 3.53 \text{Vm}^{4/3} A^{-2/3} s^{-2/3}$. As shown by Figure 5.15, this increase becomes significant for $N_b > 10^{17} \text{cm}^{-3}$. The threshold voltage is therefore increased according to (5.14).

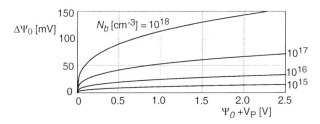

Fig. 5.15. Increase of inversion potential Ψ_0 due to band gap widening.

Furthermore, band gap widening reduces the specific current I_{spec} by approximately $1+K_cC_{ox}$, with a value of K_c between 10^{-2} and $2\times 10^{-2}\mu m^2/fF$. Except in weak inversion for which it is irrelevant, this reduction becomes significant for $C_{ox} > 5$ fF/μm^2.

5.8.4 Gate Leakage

When t_{ox} is reduced below 3nm, a gate-to-channel leakage current I_G starts to appear even at low gate voltages as the result of direct tunneling of electrons through the gate oxide. This current increases exponentially with the reduction of t_{ox} due to the increasing tunneling probability, which is also strongly dependent on the voltage across the oxide. I_G is approximately proportional to the saturated drain current I_F in weak inversion, but it increases faster than I_F in strong inversion [52].

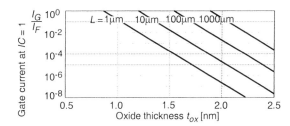

Fig. 5.16. Variation with oxide thickness of the relative gate current at $IC = 1$ (typical case).

Since $I_G \propto WL$ whereas $I_F \propto W/L$, the ratio I_G/I_F increases with L^2, as shown in Figure 5.16 for a typical case calculated at $IC = 1$ (upper limit of weak inversion). As can be seen, for $t_{ox} > 2.5$nm, I_G is negligible for most applications even for $W = 1$mm.

However, this is only the gate-to-channel leakage, which is limited by the limited number of electrons available in the channel. In practice, the total gate leakage in weak and moderate inversion might be dominated by tunneling in the gate-source and gate-drain overlap regions, especially for small values of L.

5.8.5 Drain-Induced Barrier Lowering (DIBL)

Each of the the source-substrate and drain-substrate junctions creates a barrier of potential Φ_B. For no applied voltage and in the flat-band situation ($\Psi_s = 0$) this barrier is given by

$$\Phi_B = U_T \ln \frac{N_{diff} N_b}{n_i^2}, \qquad (5.85)$$

where N_{diff} is the doping concentration of the source and drain diffusions. This barrier is increased for $V_{S,D} > 0$. When the surface potential Ψ_s is increased above zero by increasing the gate voltage, the barrier height is decreased, allowing electrons to be injected in the channel. They can then diffuse from source to drain to produce the weak inversion current.

Now, the full potential transition across these barriers occurs along some length. For a long channel, this length is negligible, and the surface potential reaches the value imposed by the gate voltage V_G, as shown by Figure 5.17. But the same figure shows that, if the channel is too short, the two barriers invade the whole channel, and Ψ_s can nowhere reach the value imposed by V_G. The maximum barrier height is reduced by an amount $\Delta\Psi_{smin}$, which will increase the drain current I_D. Moreover, $\Delta\Psi_{smin}$ increases with the drain voltage; I_D therefore depends on V_D even in saturation.

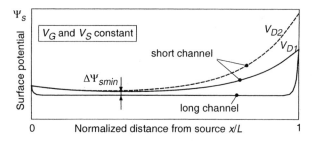

Fig. 5.17. Drain-induced barrier lowering (DIBL) by an amount $\Delta\Psi_{smin}$ in a short-channel transistor.

DIBL is mostly affecting weak inversion, since the current is then exponentially dependent on the surface potential. The reduction of the barrier height in weak inversion is given by [52]

$$\Delta\Psi_{smin} = 2e^{-\lambda/2}\sqrt{(\Phi_B - \Psi_0 - V_P + V_S)(\Phi_B - \Psi_0 - V_P + V_D)}, \qquad (5.86)$$

where

$$\lambda = \lambda_0 \left(\frac{\Psi_0}{\Psi_0 + V_P}\right)^{1/4} \quad \text{and} \quad \lambda_0 = L\sqrt{\frac{qN_b}{\epsilon_{si}\Gamma_b}}\Psi_0^{-1/4}. \tag{5.87}$$

It is exponentially dependent on L through λ_0, whereas its dependence on V_S and V_D appears inside the square root. The current in weak inversion is obtained by modifying (5.39) to

$$I_{F,R} = I_{spec} \exp \frac{V_P - V_{S,D} + \Delta\Psi_{smin}}{U_T}. \tag{5.88}$$

The resulting drain current $I_F - I_R$ is shown in Figure 5.18 for $V_P = -5U_T$ (weak inversion) and $V_P = 0$ (moderate inversion). As can be seen, the current

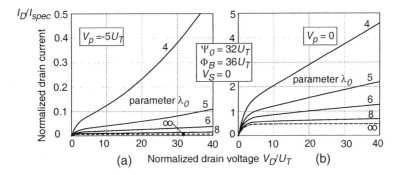

Fig. 5.18. Effect of DIBL on output characteristics: (a) example in weak inversion; (b) example in moderate inversion.

in weak inversion is dramatically increased for $\lambda_0 > 6$. The effect is still important in moderate inversion, but it is progressively reduced in strong inversion.

The effect of DIBL on transconductances in weak inversion can be obtained by differentiation of (5.88). The source transconductance becomes

$$G_{ms} = \frac{I_F}{U_T}\left(1 - \frac{d\Delta\Psi_{smin}}{dV_S}\right) = \frac{I_F}{U_T}\left(1 - e^{-\lambda/2}\sqrt{\frac{(\Phi_B - \Psi_0 - V_P + V_D)}{(\Phi_B - \Psi_0 - V_P + V_S)}}\right). \tag{5.89}$$

It is reduced by a fraction that increases exponentially with $1/L$. The gate transconductance is more complicated to express, since λ also depends on V_G through V_P, but it is reduced by about the same fraction.

Most dramatic is the fact that I_F also depends on V_D, which produces a residual drain transconductance in saturation G_{mdsat} given by

$$G_{mdsat} = \frac{I_F}{U_T} \cdot \frac{d\Delta\Psi_{smin}}{dV_D} = \frac{I_F}{U_T} \cdot e^{-\lambda/2}\sqrt{\frac{(\Phi_B - \Psi_0 - V_P + V_S)}{(\Phi_B - \Psi_0 - V_P + V_D)}}. \tag{5.90}$$

5 EKV Model of the MOS Transistor

This effect overcomes the channel shortening effect discussed in Section 5.4.2. G_{mdsat} replaces G_{ds} in the expression (5.49) of the maximum possible voltage gain, which is drastically reduced, as shown in the example of Figure 5.19.

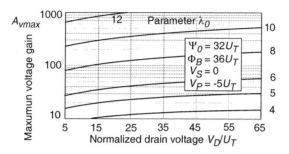

Fig. 5.19. Effect of DIBL on maximum voltage gain.

DIBL is the only important short-channel effect on a transistor operated in weak inversion. Indeed, velocity saturation is negligible, since the current is carried by diffusion, and the carriers do not reach a velocity approaching its saturation value. Therefore, a short channel does not increase noise, and the speed keeps increasing with $1/L^2$.

6

Digital Logic

Co-authored by Joyce Kwong

Digital circuits operating in the sub-threshold region provide the minimum energy solution for applications with strict energy constraints. We showed in the previous chapter that many designs exhibit a minimum energy operating point higher than the minimum achievable V_{DD}, and this operating point is a function of several parameters [94][103][99]. In general, designs with larger leakage energy relative to active energy have a higher optimum V_{DD}.

This chapter examines the operation of various digital logic cells in sub-threshold. First we analyze the CMOS inverter in detail and then extend the analysis to a CMOS sub-threshold standard cell library. After considering theoretically optimal sizing, we explore minimum energy operation for CMOS standard cell designs. A fabricated 0.18μm test chip provides measurements for analysis. We also compare the P-nMOS and transmission gate logic styles to static CMOS in the sub-threshold region in the context of process variations.

6.1 Inverter Operation in Sub-threshold

This section uses the basic static CMOS inverter, shown in Figure 6.1, to illustrate digital circuit operation in the sub-threshold region. The analysis for the sub-threshold inverter assumes that V_{DD} is below V_T by enough to ensure that the first order sub-threshold current equations in Section 4.2.1 are valid.

6.1.1 Sub-threshold Inverter Delay

The first order gate delay equation:

6 Digital Logic

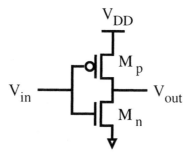

Fig. 6.1. Inverter schematic

$$t_d = \frac{C_g V_{DD}}{I_{on}} \tag{6.1}$$

provides the basis for modeling delay in both strong inversion and sub-threshold using (4.8) and (4.9), respectively. For convenience, we repeat those equations here for the case of symmetrical pMOS and nMOS transistors:

$$t_d = \frac{K C_g V_{DD}}{(V_{DD} - V_T)^\alpha} \tag{6.2}$$

$$t_{d,sub} = \frac{K C_g V_{DD}}{I_o \exp\left(\frac{V_{DD} - V_T}{n V_{th}}\right)} \tag{6.3}$$

For strong inversion operation, delay does not depend strongly on supply voltage, whereas exponentially decreasing I_{on} in the sub-threshold region leads to exponentially higher delay at lower V_{DD}. Figure 6.2 shows the normalized speed of the basic inverter across the full range of supply voltages. For $V_{DD} > V_T$, the speed decreases only slightly with voltage. In sub-threshold, the exponential roll-off of speed is very clear. This drastically reduced gate performance limits the range of applications for which sub-threshold operation is well-suited to those requiring medium or low speeds.

The lower limit for V_{DD} scaling for the sub-threshold inverter occurs when V_{DD} drops to the range of $\sim$ 3-4V_{th} [1][77]. The primary reason that the inverter functionality ceases in low voltage is degraded I_{on}/I_{off}. For strong inversion transistors operating at the nominal V_{DD}, the ratio of current for an *on* transistor ($V_{GS} = V_{DD}$), I_{on}, to its *off* current ($V_{GS} = 0$), I_{off}, is many orders of magnitude. In sub-threshold, however, I_{on}/I_{off} is greatly reduced. The rate at which this ratio decreases depends on the sub-threshold slope, S. For example, if S is 100mV per decade, then $V_{DD} = 200$mV implies that $I_{on}/I_{off} = 100$. The slower speed at lower V_{DD} results directly from lower I_{on} currents that charge and discharge circuit capacitances more slowly. Also, this degraded ratio can lead to problems with functionality at voltages considerably higher than $V_{DD} = 4V_{th}$ for some types of circuits [103].

Technologies with lower, more ideal sub-threshold slope will lose speed faster relative to strong inversion due to the faster roll-off of sub-threshold I_{on}. On the other hand, the lower S results in a larger I_{on}/I_{off} ratio for the same V_{DD}. If two technologies have the same strong inversion current, the technology with lower sub-threshold slope will be slower in sub-threshold at a given V_{DD} but will have larger I_{on}/I_{off} and a larger overall range of speed.

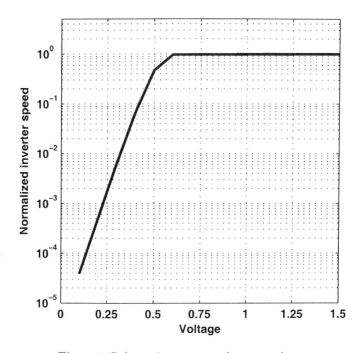

Fig. 6.2. Relative inverter speed versus voltage

6.1.2 Sub-threshold Voltage Transfer Characteristics (VTCs)

Figure 6.3 shows the I_D versus V_{DS} curves of the two transistors in Figure 6.1 arranged for load line analysis in terms of the inverter's input, V_{in}, and output, V_{out}. As V_{in} goes from 0 towards $V_{DD} = 300\text{mV}$, the nMOS curves go from bottom to top on the plot, and the pMOS curves go from top to bottom. The intersections of the lines for each value of V_{in} are marked with circles. These same circles show the points on the Voltage Transfer Characteristic (VTC) that appears in Figure 6.4. Clearly, the shape of the VTC is essentially identical to above-threshold operation. The slope of the steep part of the VTC depends on the sub-threshold slope of the inverters. A lower sub-threshold

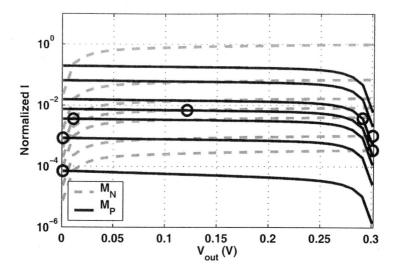

Fig. 6.3. Load line analysis at the output of a sub-threshold inverter with $V_{in} = [0.01, 0.05, 0.1, 0.125, 0.15, 0.2, 0.3]$V.

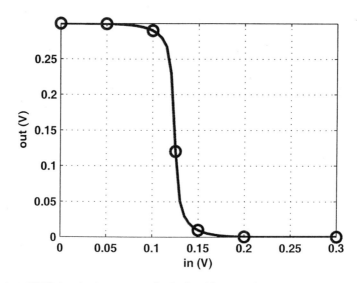

Fig. 6.4. VTC for the inverter with the load lines in Figure 6.3. $V_{DD} = 300$mV.

slope results in a steeper transition. DIBL also impacts the slope in this region. As the DIBL effect increases, the slope flattens.

As this VTC shows, the sub-threshold digital inverter behaves similarly to its strong inversion counterpart. The most notable differences for sub-threshold operation, which are clear from Figure 6.2, are the longer delays that result from lower *on*-current and the exponential dependence of I_D on V_{GS} and V_T. A good treatment of weak inversion operation for an inverter with symmetrical nMOS and pMOS appears in [53].

Inverter Sizing for Minimum Operating Voltage

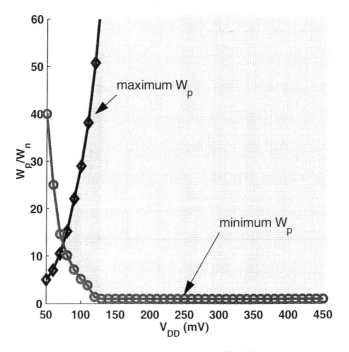

Fig. 6.5. Minimum achievable voltage retaining 10%-90% output swing for 0.18μm ring oscillator for a typical transistor (simulation).

Transistor sizing also impacts the functionality of CMOS circuits at low supply voltages. Minimum V_{DD} operation occurs when the pMOS and nMOS devices have the same current (e.g. [77]) as the following analysis will illustrate. Previous efforts have explored well biasing to match the device currents for minimum voltage operation of ring oscillators [78]. Sizing can create the same symmetry in device current.

Figure 6.5 shows the minimum voltage for which the inverter maintains 10%-90% output voltage swing. The upper bound on size occurs because the sub-threshold leakage through a large pMOS device limits the extent to which the smaller nMOS can pull down the voltage at the output. The curve denoted

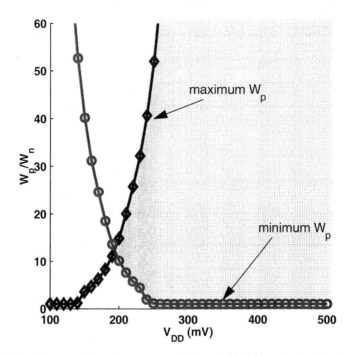

Fig. 6.6. Minimum achievable voltage retaining 10%-90% output swing for 0.18μm ring oscillator across worst case process corners (simulation). (© 2005 IEEE)

by diamonds ($W_p(max)$) shows the maximum pMOS width for which the output-low voltage of the inverter achieves 10% or less of V_{DD}. Similarly, the lower bound on size occurs because the sub-threshold leakage through a large nMOS device limits the extent to which the smaller pMOS can pull up the voltage at the output. The curve marked with circles ($W_p(min)$) shows the minimum pMOS width for which the output-high voltage achieves 90% of V_{DD}. The point where the two bounds cross in Figure 6.5 indicates the minimum operating voltage and the pMOS width needed to achieve it. For a typical transistor in this 0.18μm technology, the minimum operating voltage is 50mV, and the pMOS/nMOS sizing ratio (W_p/W_n) to achieve this minimum voltage is 12. Since minimum voltage operation occurs for symmetrical pMOS and nMOS currents, this optimum ratio tells us that the pMOS transistor in the inverter needs 12 times the width of the nMOS to equalize the sub-threshold currents in this technology.

Figure 6.5 shows theoretical operation of the inverter below 100mV, but it only applies to the typical process corner. In reality, process variations will limit operation at other corners to higher supply voltages. Therefore, analysis at the worst-case process corners is necessary to determine the minimum operating voltage for the process in general. Figure 6.6 shows another set of

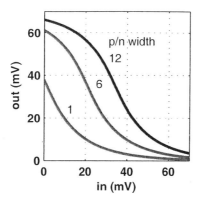

Fig. 6.7. VTCs at the minimum V_{DD} for the typical corner (simulation). pMOS/nMOS width ratio of 12 minimizes operational V_{DD} but increases energy consumption. (© 2005 IEEE)

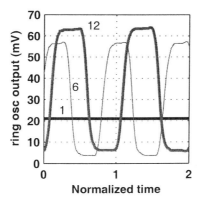

Fig. 6.8. 9-stage ring oscillator output for inverter with VTCs in Figure 6.7. (© 2005 IEEE)

min-max sizing curves for the inverter that account for the worst case process corners. $W_p(max)$ in Figure 6.6 is defined at the Weak nMOS, Strong pMOS (WS) corner, where the nMOS is much weaker than the pMOS devices to show the worst-case sizing for this condition. Likewise, $W_p(min)$ is defined at the Strong nMOS, Weak pMOS (SW) corner to provide the worst-case for pull-up. This pair of curves essentially gives the worst-case bounds for the process. Based on this analysis, the worst-case minimum supply voltage is $V_{DD} = 200$mV. Again, the pMOS/nMOS sizing ratio (W_p/W_n) to achieve operation at this voltage is 12. The ratio has not changed from typical because of the definition of the typical process corner. Since the other worst-case corners are symmetrical about the typical corner, sizing to match the pMOS and nMOS currents at the minimum corner gives the best average match in current at the global process corners as well.

Figure 6.7 shows the VTCs at a V_{DD} of 70mV for several ratios. The gain is somewhat degraded, but the optimum sized curve is symmetrical and shows good noise margins. Figure 6.8 shows the output of a 9-stage ring oscillator at the minimum voltage for the same sizes.

Widening the pMOS transistors by 12 times equalizes pMOS and nMOS currents and allows for operation at lower V_{DD}, but it also increases the switching energy consumed by the inverter. The energy savings from lowering V_{DD} are at best proportional to V_{DD}^2 if leakage is still negligible. Figure 6.6 shows that increasing pMOS widths to 12 times the nMOS width only decreases the minimum operational voltage by 60mV. This corresponds to a best-case energy savings of $0.20^2/0.26^2 = 0.6X$ due to voltage reduction. This improvement is not worthwhile if all pMOS devices are increased in size by 12X.

The exponential dependence of current in sub-threshold on V_T means that sizing is a less effective knob for 'fixing' circuits. This explains why the knob has to turn so far to match the transistor current that is imbalanced by a small difference in V_T.

6.1.3 Inverter Sizing for Minimum Energy

It has been proposed that theoretically optimal minimum energy circuits should use minimum sized devices [117], and first-order equations confirm this result for most cases.

Assuming that the majority of gates in a typical design are sized similarly, a universal increase in transistor sizes will increase both C_{eff} and W_{eff}, raising power. This type of sizing change is unlikely to decrease the critical path delay because the input to output capacitance ratios of gates will stay roughly constant, so the typical assumption of fixed capacitance loads is invalid [118]. Thus, minimum sizing also minimizes energy per operation for most generic circuits. One special case that violates this trend is a circuit with a small number of critical paths relative to the total number of paths. In this case, increased sizes on the critical path can reduce the worst-case logic depth with negligible increases in overall C_{eff} and W_{eff}, lowering E_T.

Minimum sized devices are theoretically optimal for reducing energy per operation when accounting for the impact of sizing on voltage and energy consumed [99]. Process variation in deep sub-micron processes imposes one restriction to applying this rule. The sigma for V_T variation due to random doping fluctuations is proportional to $(WL)^{-\frac{1}{2}}$, so minimum sized devices produce the worst case random V_T mismatch. Statistical analysis is necessary to confirm functionality in the face of process variation, and some devices might need to increase in size to ensure acceptable yield.

6.2 Sub-threshold CMOS Standard Cell Library

This section extends the low-voltage analysis to cells that are more complicated than the inverter. These are cells that often appear in standard cell libraries. Standard cell libraries aid digital circuit designers to reduce the design time for complex circuits through synthesis. Most standard cell libraries focus on high performance, although including low power cells is becoming more popular [118]. A lower power standard cell library generally uses smaller sizes. One standard cell library geared specifically for low power uses branch-based static logic to reduce parasitic capacitances and a reduced set of standard cells. Eliminating complicated cells with large stacks of devices and using a smaller total number of logic functions was shown to reduce power and improve performance [119]. Standard cell libraries have not been designed specifically for sub-threshold operation. This section evaluates the performance of a 0.18μm standard cell library in sub-threshold operation.

Standard Cell Library Analysis

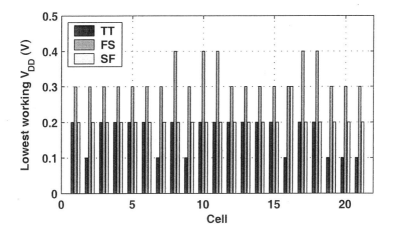

Fig. 6.9. Standard cell functionality in synthesized FIR filter using normal cell selection over process corners (simulation). (© 2005 IEEE)

In this section, we discuss the potential problems that may arise in standard cell gates that prevent them from operating at low supply voltages. We evaluate a standard cell library for sub-threshold functionality by analyzing each cell individually. The goal for this analysis is to isolate cells that fail at voltages above or near the minimum energy operating V_{DD} and to either replace or redesign those cells. If all of the cells in the library function at voltages below the optimum V_{DD}, then operation at the optimum point is guaranteed.

Figure 6.9 shows the minimum operating voltages for a subset of different standard cells in a 0.18μm library. The Typical nMOS, Typical pMOS (TT) and worst-case (SW and WS) process corners are shown. All of the cells operate at 200mV at the typical corner, showing the robustness of static CMOS logic. Additionally, most of the cells operate at 300mV in the worst case. The cells which exhibit the worst case (failing below 400mV) are logic gates with stacks of series devices (e.g. And/Or/Invert (AOI)), logic gates with multiple devices in parallel and flip-flops. More details about these problematic cells appear in the following sections.

6.2.1 Parallel Devices

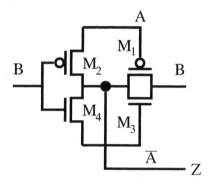

Fig. 6.10. Tiny XOR gate used in standard cell libraries. (© 2005 IEEE)

Connecting multiple devices in parallel to a single circuit node can significantly raise the minimum operating voltage in some cases. Specifically, several parallel *off* transistors effectively increase the I_{off} of the gate and degrade the I_{on}/I_{off} ratio. This degraded ratio can impact the functionality of the gate. An example of leakage in parallel devices is illustrated in the XOR gate. Figure 6.10 shows the schematic of the tiny XOR gate commonly found in standard cell libraries. In strong inversion, this gate does not exhibit any problems pulling the output voltage high at node Z. However, when the voltage scales toward 100mV, the *off* currents of the pull-down devices can dominate the *on* current of the single pullup device.

Figure 6.11(a) shows the *on* and *off* current flow diagram for the case when A=1 and B=0. M_2 acts to pull Z to the full V_{DD} voltage, but parallel leakage through M_1, M_3 and M_4 fights against M_2 and pulls the voltage at Z toward ground. At 100mV, the low I_{on}/I_{off} ratios cause the output to droop to an intermediate voltage as shown in the resulting waveform in Figure 6.11(b).

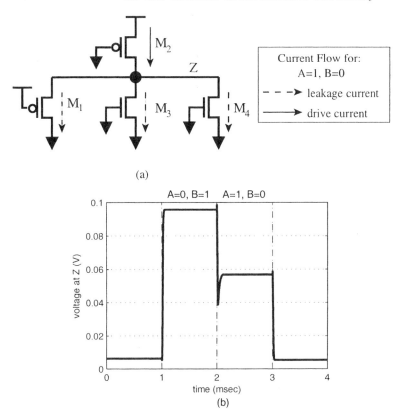

Fig. 6.11. Tiny XOR gate (a) current flow diagram for A=1, B=0 and (b) waveform. (© 2005, IEEE)

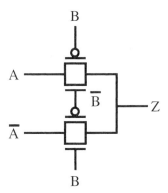

Fig. 6.12. Transmission gate XOR for sub-threshold operation. (© 2005 IEEE)

A different XOR design shown in Figure 6.12 does work at ultra-low voltages. The sub-threshold XOR uses transmission gates to minimize or balance

86   6 Digital Logic

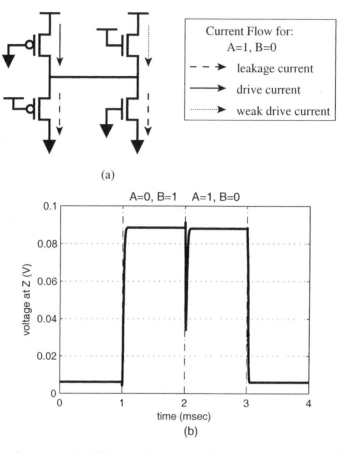

Fig. 6.13. Sub-threshold XOR gate (a) current flow diagram for A=1, B=0 and (b) waveform. (© 2005 IEEE)

the number of devices in parallel for minimum-voltage operation. The waveform in Figure 6.13 shows that this design functions for all inputs without exhibiting voltage droop at the output. Analysis of transmission gate logic with process variations is discussed in Section 6.3.

6.2.2 Stacked Devices

Stacked devices have two effects on the functionality of circuits in subthreshold operation. First, when stacked devices are conducting, the effective drive current of the two devices is diminished (e.g. the drive current of two stacked devices is approximately halved). Second, the threshold voltage in a stacked device increases when the source-to-body voltage increases. This leads to a decrease in the leakage current through the stacked device [96].

The impact of the stack effect and parallel devices is seen in a 2-input and 3-input NAND simulation. The analysis showed that the minimum-voltage operation of the 3-input NAND is 15mV higher than that of the 2-input NAND. This occurs because the effective I_{on}/I_{off} is lower for the worst case input vector combination. For the 2-input NAND, when the inputs are at '1', there are two "off" pMOS devices and two nMOS devices in a stacked configuration. For the 3-input NAND, when all inputs are at '1', there are three "off" pMOS devices leaking into the output node, and three nMOS devices stacked. The increased parallel devices cause higher I_{off} and the increased stacking causes lower I_{on}. Therefore, the minimum operating voltage increases.

6.2.3 Flip-flops

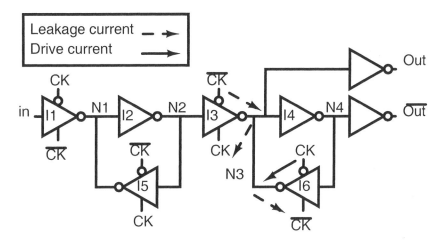

Fig. 6.14. Standard cell flip-flop at worst-case failure point where $CK = 0$ at SW corner (strong nMOS, weak pMOS). (© 2005 IEEE)

Figure 6.14 shows a schematic of the D-flip-flop commonly found in standard cell libraries. In the standard implementation, all of the inverters use small nMOS and only slightly larger pMOS devices except I3, which is several times larger to reduce Clock-to-Q delay. At the SW corner, the narrow pMOS in I6 cannot hold N3 at a one when CK is low. This is because the combined, strong *off* current in the nMOS devices in I6 and I3 (larger sized) overcomes the weakened, narrow pMOS device in I6. Tying back to the ring oscillator in Figure 6.5, the combined nMOS devices create an effective pMOS/nMOS ratio that is less than one. The original flip-flop fails to operate below 400mV at the corner. To improve this, we reduced the size of I3 and strengthened I6. Clearly, the larger feedback inverter creates some energy overhead. However, the resized flip-flop can operate at 300mV at all process corners in simulation.

6.2.4 Ratioed Circuits

Generally, static CMOS circuits are robust for sub-threshold operation. The large sensitivity of sub-threshold circuits to process variation makes some circuit designs more suitable than others. Ratioed circuits can create problems with functionality in sub-threshold.

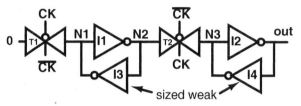

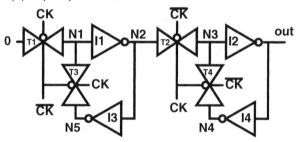

Fig. 6.15. Two flip-flops for evaluation for sub-threshold operation. Flip-flop A has ratioed feedback, and Flip-flop B does not. Ratioed circuits cannot function across process corners in sub-threshold at the minimum energy voltage because the exponential dependence of current on V_T becomes more important than sizing. Cutting the feedback loop for writing a latch is robust across all corners for operation at the minimum energy voltage. (© 2006 IEEE)

For example, Figure 6.15 demonstrates an important consideration about circuit selection for sub-threshold. The figure shows a ratioed flip-flop (Flip-flop A) and a flip-flop with transmission gates for cutting off feedback in the latches (Flip-flop B). The ratioed flip-flop uses devices sized to be strong (wide) on the critical path and inverters sized to be weak for the feedback in the latches. This approach depends on the sizing ratios in the circuit to guarantee functionality, and it works well above-threshold. In sub-threshold, however, current depends exponentially on V_T. This makes process variation and local device variation quite significant relative to device sizes. Figure 6.16 shows a plot of Flip-flop A failing to write node N3 to a one at the strong nMOS, weak pMOS corner. The pull-up path through I1 and T2 is weak because of

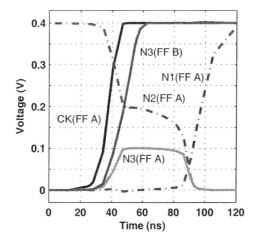

Fig. 6.16. Master stage of Flip-flop A fails to write at the strong nMOS, weak pMOS corner at 400mV and is itself overwritten. (© 2006 IEEE)

the higher pMOS threshold voltage and cannot overcome the pull-down path through I4 and T2 that is made strong by the lower nMOS threshold voltage. Instead, the pull-down path overcomes I1 and actually flips the state of the master stage by pulling N2 to zero. Thus, Flip-flop A fails below 450mV at this corner, preventing the circuit from reaching its minimum energy voltage. Massively up-sizing the pMOS devices can correct this problem, but a better choice is to eliminate the ratioed fight by adopting Flip-flop B. This flip-flop cuts off the feedback path before writing a latch, allowing the V_{GS} applied to on-transistors to increase current beyond any device off-currents even at process corners.

In general, the strong dependence of sub-threshold current on V_T and temperature makes the ratio of sizing inadequate for compensating across the full process corner and temperature ranges in sub-threshold operation. As a result, non-ratioed circuit styles provide more robust functionality in sub-threshold.

6.2.5 Measured Results from Test Chip

We use an 8-bit, 8-tap, parallel, programmable FIR filter as a benchmark to compare a normal standard cell library to a library that is modified to minimize the operational V_{DD}. The modified library is the same standard cell library but with cells that are modified to allow operation at the minimum energy point across all process corners. We eliminated the problematic cells (c.f., Figure 6.9) by preventing the synthesis tool from selecting logic gates with large device stacks and by re-sizing a few offending cells such as the flip-flop and full adder.

A 0.18μm, 6M layer, 1.8V, 7mm² test chip was fabricated to measure the impact of sizing on minimum energy operation of standard cells. The test chip

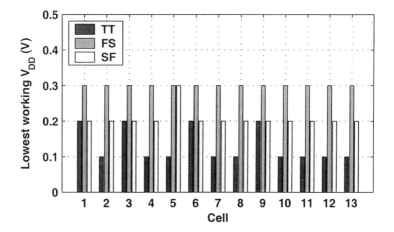

Fig. 6.17. Standard cell functionality in synthesized FIR filter using cells sized to minimize V_{DD} over process corners (simulation). (© 2005 IEEE)

features two programmable 8-bit, 8-tap FIR filters. Both filters produce non-truncated 19-bit outputs. The first filter was synthesized using the unmodified synthesis flow and normal cells (Figure 6.9). The second filter was synthesized using the modified flow in which some cells were omitted and some cells were resized to minimize V_{DD} (Figure 6.17). Both filters can operate using an external clock or an on-chip clock generated by a ring oscillator that matches the respective critical path delay of the filters. Input data comes from an off-chip source or from an on-chip Linear Feedback Shift Register (LFSR).

Figure 6.17 shows the lowest operating voltage for the cells in the minimum-V_{DD} FIR filter. The number of cell types has reduced, and all of the cells work at 300mV across all corners.

Figure 6.18 shows the measured performance versus V_{DD} for the two filters using their respective critical path ring oscillators and the LFSR data to produce one pseudorandom input per cycle. The minimum-V_{DD} filter exhibits a 10% delay penalty over the standard filter. Both filters operate in the range of 3kHz to 5MHz over V_{DD} values of 150mV to 1V. Both filters are fully functional to below 200mV.

Figure 6.19 shows an oscilloscope plot of the standard filter working correctly at $V_{DD} = 150$mV. The clock in this plot is produced by the ring oscillator on-chip. The reduced drive current and large capacitance in the output pads of the chip cause the slow rise and fall times in the clock, but the signal is still full swing. One bit of the output is shown.

Figure 6.20 shows the measured total energy per output sample of the two FIR filters versus V_{DD}. The solid line is an extrapolation of $C_{eff}V_{DD}^2$ for each filter, and the dashed lines show the measured leakage energy per cycle. Clearly, both filters exhibit an optimum supply voltage for minimizing

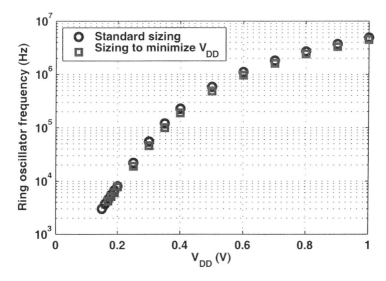

Fig. 6.18. Measured performance of programmable FIR filters on the test chip. Standard FIR is 10% faster than the minimum-voltage FIR. (© 2005 IEEE)

the total energy per cycle. Within the granularity of the measurements, the optimum V_{DD} is 250mV for the standard FIR, which matches the analytical solution derived in Section 4.2.3. The optimum V_{DD} is 300mV for the minimum-V_{DD} FIR. There is a measured overhead energy per cycle of 50% in the filter sized for minimum V_{DD}. The figure also shows the simulated worst-case minimum V_{DD} for the two filters (cf. Figure 6.9, Figure 6.17). Accounting for overhead at the worst-case minimum V_{DD}, the minimum-V_{DD} FIR offers a reduction in total energy of less than 10% at the worst-case process corner, but this improvement comes at a cost of 50% at the typical corner.

Simulations show that the measured overhead cost in the minimum-V_{DD} filter primarily results from restricting the cell set that the synthesis tool could use. Since the tool was not optimized for the smaller set of cells, we did not see the improvements that are possible through this approach [119]. Using only sizing to create the minimum V_{DD} filter would have decreased the overhead. However, the shallow nature of the optimum point in Figure 6.20 shows that the unmodified standard cell library does not use much extra energy by failing at a higher V_{DD} at the worst-case corner. Thus, existing libraries provide good solutions for sub-threshold operation. Simulation shows that a minimum-sized implementation of the FIR filter has 2X less switched capacitance than the standard FIR, so a mostly minimum-sized library theoretically would provide minimum energy circuits. Figure 6.21 shows the die photograph of the 0.18μm FIR test chip.

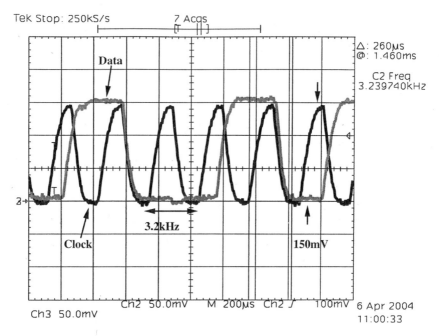

Fig. 6.19. Oscilloscope plot from the test chip showing $V_{DD} = 150\text{mV}$ filtering operation with ring oscillator clock at 3.2kHz. (© 2005 IEEE)

6.3 Logic Families in Sub-threshold

Process variation is of increasing concern in modern deep sub-micron technologies. The exponential dependence of sub-threshold currents on the threshold voltage further magnifies the impact of V_T variations. This has important implications in device sizing and choice of logic styles for sub-threshold circuits at advanced technology nodes.

Some previous work has examined different logic families for ultra-low voltage operation. Static CMOS and transmission gate designs provided a robust solution for sub-threshold test chips in [120] and [103]. It has been suggested in [121] that P-nMOS circuits demonstrate better power-delay product in simulation because of the lower delays associated with smaller load capacitances. Dynamic logic has also been suggested for sub-threshold operation [122]. This section revisits various logic families in the context of process variations.

6.3.1 Process Variation in Sub-threshold Logic

Process variations can be classified into global (inter-die) and local (intra-die) variations [123]. Global variations affect all devices on a die in the same way, for example, from wafer-to-wafer discrepancies in alignment or processing

6.3 Logic Families in Sub-threshold 93

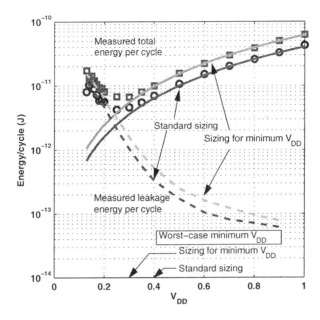

Fig. 6.20. Measured energy per operation of the FIR filters on the test chip. (© 2005 IEEE)

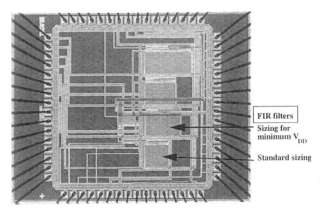

Fig. 6.21. Annotated die photo of 0.18μm sub-threshold FIR test chip. (© 2005 IEEE)

temperatures. In sub-threshold logic, the main effect of global variation is seen at skewed P/N corners with a strong pMOS and weak nMOS, or vice versa. As addressed in Section 6.1, a logic gate may not function correctly if the pull-up or pull-down network cannot drive the gate output to a full logic 0 or 1 level.

6 Digital Logic

Local variations affect devices on the same die differently and can consist of both systematic and random components. For example, an aberration in the processing equipment gives rise to systematic variation, while placement and number of dopant atoms in device channels contribute to random variation. The mismatch between devices on the same die is often modeled as a difference between their threshold voltages, with V_T being normally distributed. The standard deviation of V_T variation has been shown in [124] to be approximately proportional to $(WL)^{-\frac{1}{2}}$.

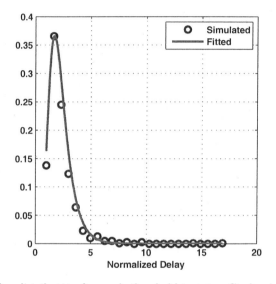

Fig. 6.22. Delay distribution for a sub-threshold inverter. Circles denote simulated data points, fitted to an ideal lognormal distribution (solid line).

This spread of V_T causes both delay and energy of sub-threshold circuits to follow a lognormal distribution, since currents have an exponential dependence on V_T. Figure 6.22 shows a typical delay distribution for a sub-threshold inverter obtained by varying the V_T of each transistor in a Monte-Carlo simulation. In addition, random V_T variation changes the I_{on}/I_{off} ratio of logic gates, resulting in a distribution of output swings. Thus local variation affects not only performance, but also functionality of sub-threshold circuits. The latter effect is quantified here by recording the DC output voltages of a logic gate having a capacitive load and buffered inputs, while random V_T mismatch is applied only to the gate under test. Figure 6.23(a) and (b) show a 1k-point Monte-Carlo simulation of the output-low (V_{OL}) and output-high (V_{OH}) levels of a minimum sized static CMOS inverter. The majority of the gates in the simulation function as expected, but several outliers exist (marked

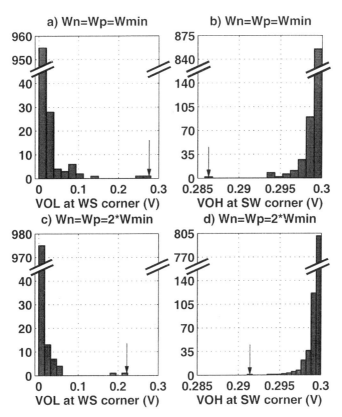

Fig. 6.23. Histogram of output levels in a static CMOS inverter when $V_{DD}=300$mV. Arrows point to extreme outliers. V_{OL} and V_{OH} are measured at the weak-nMOS, strong-pMOS and strong-nMOS, weak-pMOS global corners respectively.

by arrows). In Figure 6.23(c) and (d), an inverter with devices of two times minimum width displays smaller variability in output levels. Clearly, functionality is a major concern in sizing sub-threshold logic when accounting for local variations.

6.3.2 Evaluating Logic Styles in the Context of Variations

Variations also affect the choice of logic styles. Section 6.2 discussed how certain circuit topologies, for example a ratioed circuit or one with many parallel devices, are less robust to global variations. This section examines the additional impact of local variation on several logic families.

In order to obtain a fair comparison, circuits of different logic families should be sized according to a consistent metric. Section 6.2 has shown that minimum size devices are theoretically optimal for minimizing energy. However, when local variations are significant, some devices must be made larger in order to achieve the required robustness, at the expense of higher energy consumption. The following discussion first presents a sizing metric with robustness as the primary goal, then compares 8-bit adders implemented in each logic style in terms of energy, delay, and variability. Simulations in this section are performed in an industrial 65nm process to accurately capture the impact of random V_T variation on sub-threshold circuits.

Sizing Metric

It was previously observed that both global and local variation affect the output voltage of sub-threshold logic gates. In this logic style comparison, devices are upsized to achieve acceptable V_{OL} and V_{OH} for proper functionality at a minimum supply voltage of 300mV. The requirements are set at 20% to 80% of V_{DD} respectively.

For each logic style, we first size transistors considering global variation alone according to the sizing curves of section 6.1. We then add the effect of local variation and perform Monte-Carlo simulation varying V_T of all transistors at global process and temperature corners. Device widths are increased until 99.9% of samples have satisfactory output voltage.

The resulting sizing trends for each logic style are shown in Table 6.1 and discussed in detail below. Values are normalized to the minimum width in the process. If only global variation is considered (left column), minimum size devices are sufficient to achieve the required voltage swing at 300mV. In the presence of local variation, transistors must be upsized to mitigate variability in output levels.

Static CMOS Logic

In static CMOS logic, we consider circuit primitives consisting of one, two, and three pMOS and nMOS devices in series. When considering global variation alone, minimum size devices in both pull-up and pull-down networks are sufficient to produce the required output swing at $V_{DD} = 300$mV. When local V_T variation is taken into account, stacks of devices display higher variability in output levels. As seen in Table 6.1, series devices must be upsized to mitigate this effect to achieve the desired 99.9% yield in terms of output swing.

Transmission Gate Logic

Transmission gate logic is sized similarly to static CMOS, by measuring a gate's output swing with variation and increasing device sizes accordingly. Since this logic style often involves several transmission gates connected to

6.3 Logic Families in Sub-threshold

Table 6.1. Variation-driven device sizing for circuit primitives.

Circuit primitives	Min. size constraint	
	Global variation	Global and local variation
Static CMOS		
1 nMOS in pull-down	1	2.67
2 series nMOS in pull-down	1	5.33
3 series nMOS in pull-down	1	6.33
1 pMOS in pull-up	1	1
2 series pMOS in pull-up	1	1
3 series pMOS in pull-up	1	1.67
Transmission gate		
Transmission gate	1	2
Pseudo-nMOS		
2 series nMOS in pull-down	1	5.83
3 series nMOS in pull-down	1	7.00

the same output node, idle leakage through gates in the off state can degrade the output level of the gate in the on state. Local variation may further weaken the active current relative to leakage currents such that the output level lies outside specification. This is accounted for in simulation by including leakage through three off gates to fight the gate under test, as shown in Figure 6.24.

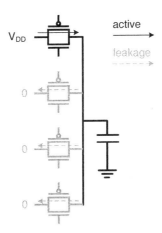

Fig. 6.24. Sizing of transmission gate (black) while accounting for leakage currents in off gates (gray).

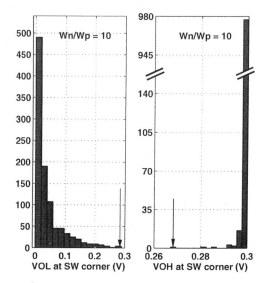

Fig. 6.25. Histogram of output levels in P-nMOS inverter at V_{DD}=300mV and best-case, strong-N weak-P corner. Arrows point to extreme outliers.

Pseudo-NMOS (P-nMOS) Logic

[121] first proposed the use of P-nMOS logic in sub-threshold. It was reported that a P-nMOS inverter in a 0.35μm process displayed better voltage transfer characteristics in sub-threshold than in above-threshold, with lower V_{OL} and higher gain in the transition region. This was due to a flat quasi-saturation region and steep roll-off in the quasi-linear region of the sub-threshold I_D versus V_{DS} curve. Basic P-nMOS logic gates also exhibited lower delays from reduced load capacitances compared to their static CMOS counterparts. The additional power from static current was found to be relatively small such that P-nMOS gates had lower power delay product (PDP) than static CMOS in sub-threshold.

Sub-threshold P-nMOS is revisited for the 65nm process of this section. The observed VTC at low supply voltages is degraded by a sloping quasi-saturation region caused by DIBL. Furthermore, when total energy per cycle rather than PDP is the metric of interest, P-nMOS becomes less desirable because static currents contribute significantly to leakage energy when they are integrated over long critical path delays in sub-threshold.

When local variation is additionally taken into account, P-nMOS logic generally displays higher variability in output levels than the other two logic styles. In P-nMOS, the n-type MOSFET (nMOS) is fighting a p-type MOSFET (pMOS) that is always on, so a small increase in V_T of the pull-down network translates into a large increase in V_{OL}. Figure 6.25 plots the output level distribution of a P-nMOS inverter with a minimum pMOS width

6.3 Logic Families in Sub-threshold

and a ratio W_N/W_P of 10. The simulation is performed at the strong-nMOS weak-pMOS corner, assuming that substrate biasing is available to shift relative P/N strengths of all dies to the best-case condition. Even with efforts to strengthen nMOS relative to pMOS by increasing the pMOS length above minimum and operating at the best-case skewed corner, the distribution shows a much higher deviation compared to a static CMOS inverter.

Because of the large variability, there is a significant trade-off between high functional yield (requiring large nMOS) or low delay and energy (small nMOS) in sizing P-nMOS logic. The ensuing discussion applies the latter option by sizing P-nMOS gates for an input capacitance comparable to their static CMOS counterparts. For larger sizes of the nMOS gates, P-nMOS logic no longer offers the benefit of lower load capacitance.

Dynamic Logic

For strong inversion circuits, dynamic logic provides high-speed operation but is less desirable in low-power design because of additional overhead. There are further disadvantages of using dynamic logic in sub-threshold. At a low supply voltage, there is only a small amount of charge stored on the dynamic node. The dynamic node becomes highly susceptible to noise and idle leakage currents. The charge leakage problem worsens when variation is taken into account. When considering operation at global process corners alone, the leakage in this process does not cause the gate output to discharge to a level that switches the next logic circuit. However, if local variation also weakens the precharge device and strengthens the evaluate network, the dynamic node loses charge quickly. Since the evaluate period must be longer than the critical path delay, which has a large spread in sub-threshold, it is likely that some dynamic nodes will not be able to hold sufficient charge until the next precharge period. Because of this lack of robustness, dynamic logic is not included in the subsequent comparison.

Logic Style Comparison Results

In order to evaluate various logic styles in a circuit context, 8-bit ripple carry adders in each family [125] – mirror (static CMOS), Transmission Gate (TG) logic and Pseudo-NMOS (P-nMOS) – are designed using sizes according to Table 6.1. They are then simulated for functionality, delay, and energy.

Figures 6.26, 6.27, and 6.28 plot the probability distribution functions (PDFs) of energy and delay per addition from a 1k-point Monte-Carlo simulation of the three adders, with random V_T mismatch applied to all transistors. The same set of random input data vectors are used for all three circuits. Delay values are normalized to the Fan-Out of 4 (FO4) inverter delay and energy is measured using a system frequency of 100kHz. All measurements are taken at V_{DD}=300mV and under nominal process and temperature conditions except those for the P-nMOS adder, where a best-case, strong-nMOS weak-pMOS corner is assumed to prevent non-functionality. PDFs of the first two implementations follow the shape of a lognormal distribution. Trends in the P-nMOS adder distribution are less clear because only a small number of data points are available from the Monte-Carlo simulation. The remaining instances did not function properly and were discarded.

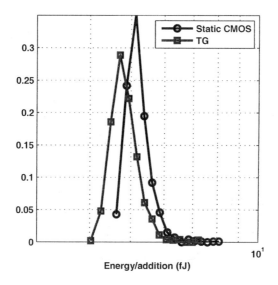

Fig. 6.26. Energy distribution of static CMOS and transmission gate adders.

Table 6.2. Number of logic failures and performance variability of three adder implementations.

	Static CMOS	TG	P-nMOS
Number of failed adders	0	0	971
Delay variability (σ/μ)	0.167	0.179	0.498
Mean (normalized) delay	18.4	17.7	15.8
Energy variability (σ/μ)	0.0718	0.0798	0.129
Mean energy [fJ]	5.22	4.87	631

Table 6.2 lists the variability and mean of delay and energy, in addition to the number of failed adders that display incorrect output data when the threshold between logic 0 and 1 is defined as $V_{DD}/2$. The TG adder offers the smallest mean energy since it avoids the stacks of large devices in static CMOS. The P-nMOS adder has the smallest mean delay, as expected from

6.3 Logic Families in Sub-threshold

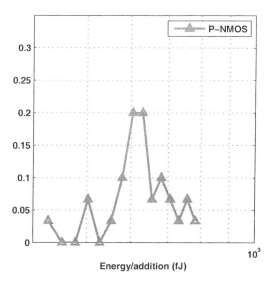

Fig. 6.27. Energy distribution of Pseudo-nMOS adder.

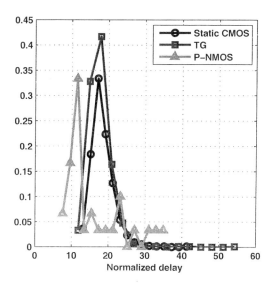

Fig. 6.28. Delay distribution for adders of three different logic styles.

reduced load capacitances. However its functional yield is very poor even at the best-case process corner, and its mean energy is two orders of magnitude higher than the other two implementations. As clock frequencies decrease exponentially with supply voltage, the total accumulated static leakage energy of idle P-nMOS circuits becomes much higher than their CMOS counterparts in deep sub-threshold operation. Due to the large energy consumption and functional failures, P-nMOS logic does not make sense for low voltage circuits in the presence of local device variations.

Large stacked devices in the static CMOS adder result in the lowest variability, while that of the TG adder is slightly higher. The P-nMOS adder displays the highest σ/μ, consistent with previous observations. It should be noted that generic transmission gate logic with chains of more than one transmission gate will show more variability. However, it is possible to realize a variety of logic functions (e.g. adder, XOR, MUX) without resorting to a chain. Therefore, Table 6.2 suggests that transmission gate-based logic in the sub-threshold region offers the benefit of robustness against local variations with relatively small device sizes, which also translate into energy savings.

7
Sub-threshold Memories

Current microprocessor and digital signal processors heavily utilize register files and memories for high speed computing. Register files are needed for local buffering and load-store operations. SRAMs providing local and global caches are starting to occupy a significant fraction of microprocessor chip area. For example, in the state-of-the-art Texas Instruments C64x Very-Long Instruction Word (VLIW) DSP processor, there is more than 1MB of memory and two 32-entry 32-bit register files [126]. The latest OMAP2 application processor contains more than 5Mb of internal RAM [127]. As processor complexity increases and transistor widths decrease, Systems on a Chip (SoCs) are adding more and more on-chip memory to the design.

In deep sub-micron technologies, large memories can dissipate a lot of leakage energy because they cannot be shutdown when idling. The information in the memories must be held. With increasing process variations in deep sub-micron, the 6T SRAM bitcell cannot function reliably at low voltages. Therefore, during idle modes the leakage of the memories at higher voltages dominates and impacts the system power dissipation. Sub-threshold operation and retention is ideal for reducing the idle mode leakage and active energy of memories for energy-constrained applications.

In this chapter, we will analyze conventional register file and SRAM designs for operation in the sub-threshold region. We will also cover new designs that allow the supply voltage of these embedded memories to scale down to sub-threshold.

7.1 Register Files

Register files are an important part of any load-store processor or DSP. As the trend goes towards wider instruction issue processors and techniques such as register renaming for speed, there is a corresponding increase in register file sizes. Because register files are located within the datapath, they are not restricted to the small SRAM pitch and can be larger in size. Also, they have higher noise immunity than a typical SRAM bitcell if designed from CMOS circuits.

A typical register file design contains 6T bitcells with multiple read and write ports. In a register file, energy is dissipated in the bitlines, the wordlines, the sensing schemes and the control logic. Various power saving techniques have been used such as low-voltage swing drivers and bitline partitioning [128] for improved leakage and active power.

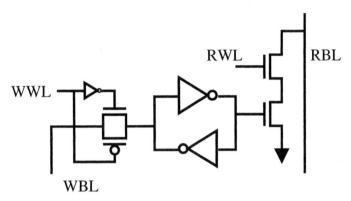

Fig. 7.1. Strong inversion multiported RAM (© 2005 IEEE)

Figure 7.1 shows a single-ended register file configuration used for strong inversion operation [129]. The main cell in the register file array is the cross coupled inverter pair. A transmission gate is used to write to the memory bit and an nMOS pull-down is used to read from the memory bit. This register file design is suitable for strong inversion operation, but at low-voltage it is not possible to write to the bitcell.

7.1.1 Write Port and Memory Cell

The write port of the sub-threshold register file has two parts. The first part is the decoding logic which decodes the write address and asserts the correct wordline. The wordline signal (WWL) drives the input to the register. The second part consists of the write mechanism to pass the data into the memory cell.

Replacing the strong inversion write port to the memory cell design with a C^2MOS tristate gate enables the write operation to occur into the sub-threshold region. Also, the feedback inverter inside the memory bitcell itself is changed to a tristate inverter. The resulting bitcell, shown in Figure 7.2, achieves 215mV operation for write accesses across process corners in simulation, which is a much lower voltage than the ratioed circuit showed in Figure 7.1.

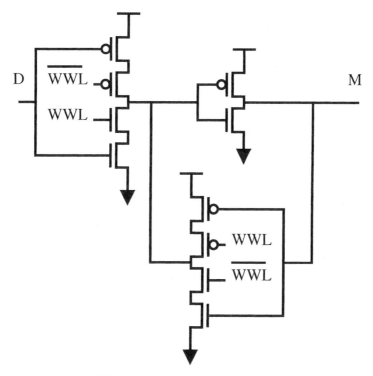

Fig. 7.2. Latch based write port

7.1.2 Read Bitline Architectures

Conventional register file read-ports use sense-amplifier circuits which detect a small bitline differential and quickly restore rail-to-rail outputs. Traditionally sense-amplifiers are triggered with a 100mV difference between the bitlines and are used to improve read access times.

In sub-threshold operation, the sense-amplifier design becomes extremely difficult. As the voltage supply drops down towards 100mV, the voltage differential of the sense amplifier is on the order of 10's of millivolts. The tiny differential and effect of noise on dynamic nodes makes the sense-amplifier design highly unreliable at low voltages. In addition, since clock speed is not the key metric for sub-threshold applications then sense-amplifier based read bitlines are not required and large signal sensing is preferred in sub-threshold because it is more robust at low-voltage.

In [130], the authors showed that small signal sensing did have a good delay area trade-off for $0.18\mu m$ and $0.13\mu m$ technology memory design. However beyond the $0.13\mu m$ process technology, large signal sensing had a clear advantage in delay and area because the sense amplifier scheme does not scale well into deep sub-micron technologies.

7 Sub-threshold Memories

The following sections investigate different single-ended approaches to reading the register file that potentially can enable sub-threshold operation.

Precharge Pull-down Read Port

The read-only pull-down port is a scheme often found in conventional register file and memory design (Figure 7.3). During the precharge phase, the bitline is precharged to V_{DD}. During the evaluate phase, the accessed read port is enabled by a row-wordline (RWL) going high. If a '1' is stored in the memory cell then the bitline is pulled down to ground. Otherwise the bitline is held high at V_{DD}. This scheme is widely used because the bit is not disturbed during a read, and bit disturb is not worsened with multiple read-ports.

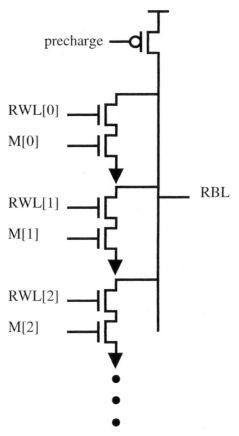

Fig. 7.3. Precharge bitline scheme. (© 2005 IEEE)

Dynamic circuits such as the precharged bitline are difficult to operate at sub-threshold voltages due to the wide range of process variations. In deep sub-micron technologies, it is difficult to keep the precharged node at V_{DD} during the evaluation phase because there is a significant amount of leakage. For fast memories, this is not a problem, since the bitline discharge is relatively slow, and the bitlines are precharged every cycle. However, the slow clock speed of sub-threshold circuits (10kHz-1MHz) causes enough time to elapse and internal dynamic nodes to drop due to leakage. The dynamic node does not discharge to ground, but rather to an intermediate voltage between V_{DD} and ground. The intermediate voltage level depends on many factors such as the number of pull-down leakage devices, the size of the precharge device, the value stored in memory and process variation.

For example, in Figure 7.3, during the evaluation phase, '1' is being driven on the bitline. The worst case output-high (RBL=1) leakage occurs when one memory bit is storing a '0' and the remaining bits store '1'. This configuration is the worst case for maintaining output-high because the worst case leakage occurs through the pulldown paths. All of the nMOS pulldown paths are single stacked because the input to the lower device is '1'. A single stacked device has higher leakage than two or more stacked devices. Thus a resistive divider is created from the leakage through the precharge device and the leakage through all of the pulldown devices. To maintain an output-high to be at least 90% of V_{DD}, a large precharge device width (W_{pre}) is required.

The worst case output-low (RBL=0) occurs for the opposite memory configuration: when one memory bit stores a '0' and the other memory bits store '1'. This configuration has the smallest amount of leakage through the pull-down paths. All but one pulldown path have stacked devices, and the body effect leads to lower leakage currents. This in itself is not a problem, but, when coupled with a large W_{pre}, the total pull-down *on* current is small compared to the pull-up precharge device. Figure 7.4 shows a waveform of the effect of bitline leakage for a 128-bit pull-down bitline operating at 100mV. The waveform shows that due to the problems of using precharge and dynamic nodes, it is not possible to differentiate between output-high and output-low. The ΔRBL is only 2mV, which is too small to detect for a successful read access.

Figure 7.5 shows a simulation to find the minimum voltage for the pull-down read bitline that gives suitable voltage differential at the output. The minimum precharge device width ($W_p(min)$) that maintains output-high to be 90% of V_{DD} is plotted versus the supply voltage. Similarly, the maximum precharge device width ($W_p(max)$) that maintains output-low below 10% of V_{DD} is plotted versus V_{DD}. The crossover point indicates the minimum operating voltage, and the shaded area indicates precharge widths and voltages which satisfy both output-high and -low requirements. The results show that the minimum operating voltage occurs at 215mV for the typical transistor corner. When incorporating worst case corners, the minimum voltage supply increases to 230mV. A simulation of a 128-bit read-bitline in 0.18μm technology shows the minimum voltage operation at 230mV with a W_p/W_n of 2900

108   7 Sub-threshold Memories

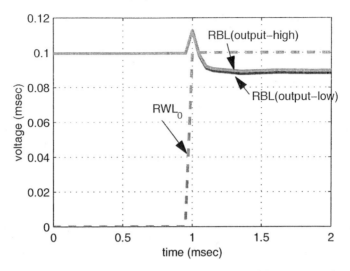

Fig. 7.4. Precharge bitline scheme waveform. (© 2005 IEEE)

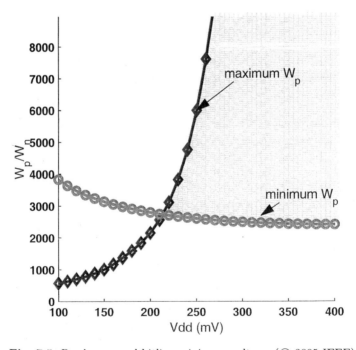

Fig. 7.5. Precharge read bitline minimum voltage. (© 2005 IEEE)

as shown in Figure 7.5. This unreasonably large sizing ratio motivates the search for better strategies for reading the register file.

Pseudo-nMOS Pull-down Read Port

Using a Pseudo-NMOS (P-nMOS) pull-up device instead of a precharge scheme helps to reduce the size of the precharge device (Figure 7.6). The gate to the pMOS is tied to ground instead of a precharge signal.

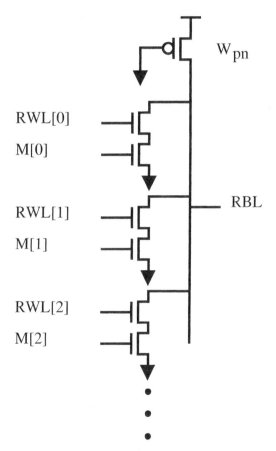

Fig. 7.6. Pseudo-nMOS read bitline minimum voltage. (© 2005 IEEE)

The minimum voltage analysis in Figure 7.7 shows that the ratio of the pseudo-nMOS device width to the minimum nMOS width (W_{pn}/W_n) is 6.7. This ratio is much smaller than that of the precharge scheme. The minimum voltage for functionality occurs at 230mV for output-low and -high swing of 10%-90% V_{DD} for typical transistors. Also, the simulation for $W_{pn}(max)$ shows a very different trend as a function of V_{DD}. As the voltage supply increases, the gate-to-source voltage of the P-nMOS device also increases, therefore reducing the need for a large W_{pn}. This is typically true for P-nMOS circuits which at high voltage only require a very weak turn-on pMOS transistor.

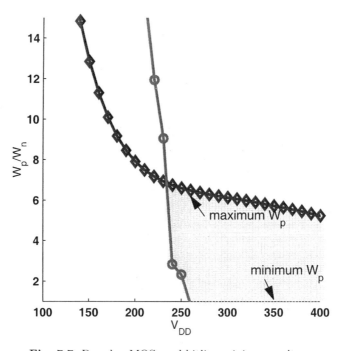

Fig. 7.7. Pseudo-nMOS read bitline minimum voltage.

Negative Wordlines

One common approach that improves the I_{on}/I_{off} of the bitlines is to apply a negative voltage to the gate of wordlines that are not accessed. The resulting negative V_{GS} significantly reduces the leakage current into the unaccessed cells. Figure 7.8 shows the negative boosted wordline scheme using negative charge pump circuits to bring the voltage from zero to a negative V_{SS} voltage. For V_{SS}=-250mV, the nMOS leakage current is up to 50X lower, which helps to mitigate bitline leakage effects.

Because the negative wordlines have lower leakage currents, a smaller precharge device is required to maintain an output-high at 90% of V_{DD}. For

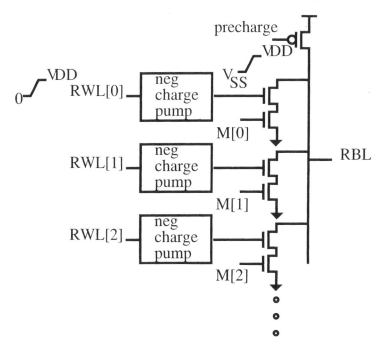

Fig. 7.8. Negative wordlines for unaccessed bitcells

the worst-case output-low, RBL is driven down to 61mV for the 128-bit example (Figure 7.9). This improved bit-line differential is achieved at the expense of the overhead required to build a negative charge pump. Also in deep submicron technology, Gate-Induced Drain Leakage (GIDL) effects appear. GIDL results when a small or negative voltage on the gate causes high transverse and lateral electric fields, leading to an increase in the drain current. The GIDL effect at large negative voltages may cancel out any benefit gained by lowering the gate voltage significantly.

Tristate Read

In sub-threshold design, CMOS circuits perform more robustly compared to dynamic circuits. Figure 7.10 shows a fully CMOS read-bitline using tristate buffers that present a high impedance to the bitline for unaccessed cells.

The tristate buffer stand-alone simulation shows full functionality down to 100mV. However, when the tristates are connected together as in Figure 7.10, the read-bitline suffers from a low I_{on}/I_{off} ratio. The worst bitline leakage for an ouptut-high occurs when one memory bit stores a '1' and the rest store '0'. When this occurs, then all except one buffer have pMOS pull-down leakage, and only one nMOS is pulling down. In sub-threshold, when I_{on}/I_{off} is on the

112   7 Sub-threshold Memories

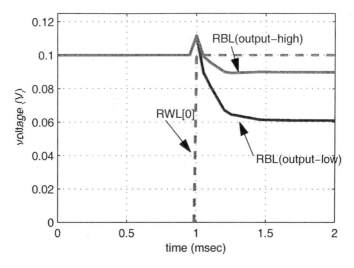

Fig. 7.9. Negative Boosted Wordline Bitline Waveform

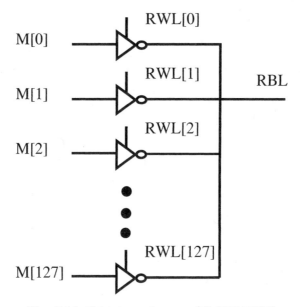

Fig. 7.10. Tristate-read access. (© 2005 IEEE)

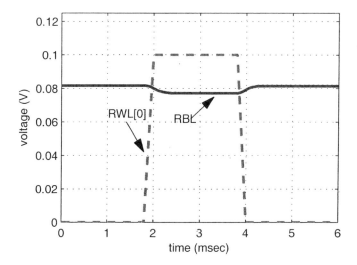

Fig. 7.11. Tristate-read access worst-case waveform. (© 2005 IEEE)

order of 10-20, then for large memories the pMOS off-current overpowers the nMOS on-current. Figure 7.11 shows the worst case input vector that causes the read-bit line to pull-down only by a few millivolts for 128-bits on the bitline. As the number of bits is reduced on the bitline, then the worst-case bitline separation improves.

Hierarchical-read Access

Another CMOS read-bitline scheme improves the I_{on}/I_{off} of the read-bitline by hierarchically breaking down the number of bits on the bitline (Figure 7.12). The read-bitline is segmented by using a 2-to-1 mux-based approach. The selectors to the muxes are the read-address inputs, and the data from the memories is hierarchically passed down through the muxes to the output. The mux has two transmission gates and an inverter and is shown in Figure 7.13. In order to avoid stacks of transmission gates and sneak leakage paths, inverters are inserted between each level of hierarchy. This scheme enables scalability to low voltages and works across process variations.

Figure 7.14 shows a waveform of the RBL for the worst case leakage. By creating a hierarchical bitline, the effect of parallel leakage for each level of hierarchy is reduced, and the design is less affected by process variations.

This read-bitline scheme is daisy-chained and arrayed as seen in Figure 7.15. This layout scheme allows for a highly compact bitline. Additionally, this scheme saves one read-address decoder because the read address is routed vertically and connected directly to the mux selector inputs.

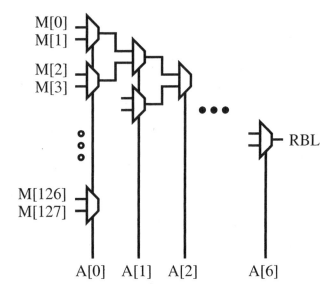

Fig. 7.12. Hierarchical-read access bitline. (© 2005 IEEE)

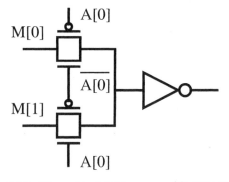

Fig. 7.13. Hierarchical Bitline mux. (© 2005 IEEE)

7.1.3 Sub-threshold Register File

A 128W x 16-bit sub-threshold register file was implemented in 0.18μm technology. In simulation, the register file operates down to 100mV for typical transistors and is sized to operate down to 300mV across worst case process corners. The register-file chip was experimentally verified with operation down to 180mV as part of a sub-threshold FFT implementation. The sub-threshold register file is the lowest operating voltage register file ever demonstrated. Section 9.1.4 gives additional details about the FFT testchip.

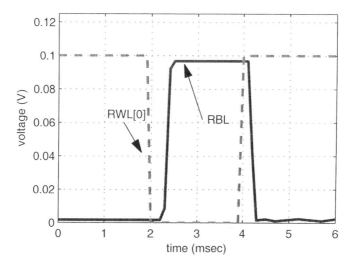

Fig. 7.14. Hierarchical-read access bitline worst-case output. (© 2005 IEEE)

7.2 Sub-threshold SRAM

The large fraction of chip area often devoted to Static Random Access Memory (SRAM) makes low power SRAM design very important as well. SRAMs comprise a significant percentage of the total area for many digital chips as well as the total power [131][132]. For this reason, SRAM leakage can dominate the total leakage of the chip, and large switched capacitances in the bitlines and wordlines makes SRAM accesses costly in terms of energy. For optimal system integration, SRAMs must operate at sub-threshold voltages that are compatible with sub-threshold combinational logic. Scaling supply voltage for SRAMs has the additional benefit of decreasing their leakage power and active energy. Recent low power memories show a trend of lower voltages with some designs holding state on the edge of the sub-threshold region (e.g. [133]). This scaling promises to continue, leading to sub-threshold storage modes and even sub-threshold operation for SRAMs operating in tandem with sub-threshold logic.

This chapter describes the investigation of an SRAM capable of operating in the sub-threshold region. Section 7.2.2 describes several key problems that prevent traditional six transistor (6T) bitcells from functioning properly in sub-threshold in a 65nm bulk CMOS process. Section 7.2.6 shows the bitcell that we developed to overcome these challenges to sub-threshold functionality, and Section 7.2.7 provides results from a 256kb SRAM test chip that uses the new bitcell.

116   7 Sub-threshold Memories

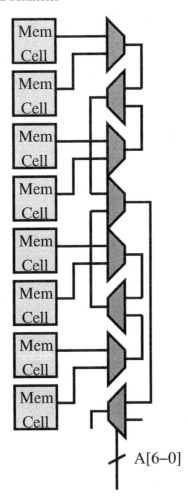

Fig. 7.15. Daisy chaining and arraying minimizes read-bitline area

7.2.1 SRAM Overview

To provide background for understanding previous low power techniques, this section gives a brief overview of the traditional 6T SRAM bitcell and its operation in the above-threshold voltage region. Despite numerous attempts to improve upon it, the 6T cell has remained the bitcell of choice for SRAM designs because of its relatively wide noise margins [134]. Figure 7.16 shows a schematic for the basic 6T bitcell. This bitcell essentially consists of back-to-back inverters that store the cell state (M_1, M_3, M_4, and M_6 in Figure 7.16) and access transistors for reading and writing (M_2 and M_5 in Figure 7.16).

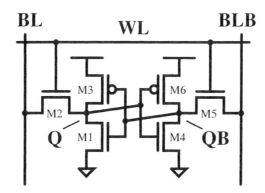

Fig. 7.16. Schematic for a standard 6T bitcell. (© 2006 IEEE)

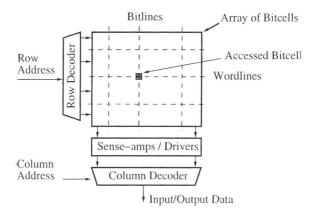

Fig. 7.17. Standard architecture for an SRAM using the 6T bitcell.

Figure 7.17 gives a simple example of a memory architecture that uses the 6T SRAM [125]. An array of bitcells is arranged into rows and columns. Each row lies along a wordline (WL), and each column is associated with a bitline (BL) pair. The memory address is divided into a row address and a column address. Decoder circuits use the applied address to select the correct wordline and bitline pair for a memory access. For a write access, the wordline is asserted (goes to '1') to turn on the access transistors (M_2 and M_5). The bitlines (BL and BLB) are driven to the correct differential value to write into the cell. The bitline driving a '0' will overwrite the data held by the cross-coupled inverters assuming that the bitcell is sized correctly [125]. For a read access, the bitlines are precharged to '1', then the wordline is asserted at the same time as the bitlines are allowed to float. The internal node of the bitcell that holds a '0' will pull its bitline low through the access transistor. Usually, a sense amplifier will detect this differential voltage on the bitlines before it becomes very large and amplify it to full voltage values. Sense amps are used primarily to speed up the read process or to avoid the energy overhead of fully discharging the large capacitance of the bitlines.

When the bitcell is holding data, its wordline is low, so M_2 and M_5 are off. In order to hold its data properly, the back-to-back inverters must maintain bi-stable operating points. The best measure of the ability of these inverters to maintain their state is the bitcell's Static Noise Margin (SNM) [135]. The SNM is the maximum amount of voltage noise that can be introduced at the outputs of the two inverters such that the cell retains its data. SNM quantifies the amount of voltage noise required at the internal nodes of a bitcell to flip the cell's contents.

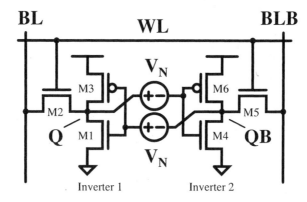

Fig. 7.18. Schematic for 6T bitcell showing voltage noise sources for finding SNM [135]. (© 2005 IEEE)

Figure 7.18 shows a conceptual setup for modeling SNM [135]. Noise sources having value V_N are introduced at each of the internal nodes in the bitcell. As V_N increases, the stability of the cell changes. Figure 7.19 shows the most common way of representing the SNM graphically for a bitcell holding data. The figure plots the VTC of Inverter 2 from Figure 7.18 and the inverse VTC from Inverter 1. The resulting two-lobed curve is called a "butterfly curve" and is used to determine the SNM. The SNM is defined as the length of the side of the largest square that can be embedded inside the lobes of the butterfly curve [135]. To understand why this definition holds, consider the case when the value of V_N increases from 0. On the plot, this causes the VTC^{-1} for Inverter 1 in the figure to move downward and the VTC for Inverter 2 to move to the right. As V_N increases, the metastable point moves closer to one of the stable points in the plot (the lower-right point in this example). Once both curves move by the SNM value, the metastable point becomes coincident with one stable point, and the curves meet at only two points. Any further noise flips the cell.

7.2 Sub-threshold SRAM   119

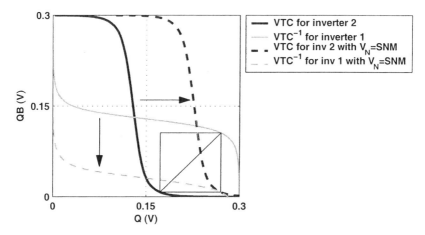

Fig. 7.19. The length of the side of the largest embedded square in the butterfly curve is the SNM. When both curves move by more than this amount (e.g. V_N=SNM), then the bitcell is mono-stable, losing its data.

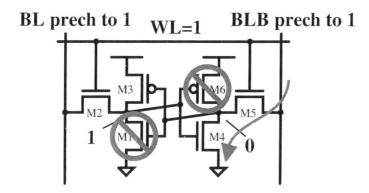

Fig. 7.20. Schematic of the 6T bitcell at the onset of a read access. WL has just gone high, and both BLs are precharged to V_{DD}. The voltage dividing effect across M_4 and M_5 pulls up node Q_B, which should be 0V, and degrades the SNM.

Although the SNM is certainly important during hold, cell stability during active operation represents a more significant limitation to SRAM operation. Specifically, at the onset of a read access, the wordline is '1' and the bitlines are still precharged to '1' as Figure 7.20 illustrates. The internal node of the bitcell that represents a zero gets pulled upward through the access transistor due to the voltage dividing effect across the access transistor (M_2,M_5) and drive transistor (M_1,M_4). This increase in voltage severely degrades the SNM during the read operation (read SNM). Figure 7.21 shows example butterfly

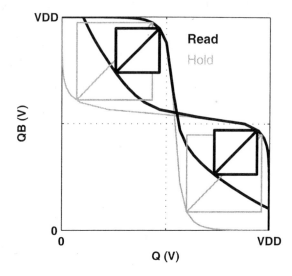

Fig. 7.21. Example butterfly curve plots for SNM during hold and read. (© 2006 IEEE)

curves during hold and read that illustrate the degradation in SNM during read. The voltage dividing effect causes the lower half of the VTC for each inverter (when its V_{in} is high) to pull upwards relative to its original position, squashing the lobes of the butterfly curve.

The following discussions of previous work deal with variations to this standard 6T bitcell and architecture. The focus of most of the works we report is to lower power, and many of the techniques can be combined with a sub-threshold SRAM design.

Voltage Scaling

Scaling of the supply voltage and other voltages related to SRAM operation has become a popular method for improving on the basic architecture. Table 7.1 shows a sampling of the different methods of voltage scaling applied to SRAM operation.

The significance of SRAM power has produced a trend of memory design aimed at lower voltage operation. Exploiting DVS for SRAM is one motivation for designing a voltage-scalable memory. A 0.18μm 32kB four-way associative cache offers DVS compatibility from 120MHz, 1.7mW at 0.65V to 1.04GHz, 530mW at 2V [149]. This memory uses high V_T in the bitcell array and low V_T transistors in the peripheral circuits. Since lowering V_{DD} amplifies the difference in delays between these two V_T regions, the architecture uses dummy bitlines to adapt the timing correctly with scaling. The bitcell itself is litho-

Table 7.1. Voltage scaling approaches used for SRAM.

Voltage	Approach	Source(s)
bitcell V_{DD}	lower in standby	[132][136][137][133][138]
	raise always	[139][140]
	raise for read access	[133][141]
	float or lower for write	[133][142]
	float for read access	[142]
	raise in standby	[143]
bitcell V_{SS}	raise in standby	[143][144][145][146][131][147][138]
	raise or float for write access	[131][148]
	lower for read access	[141]
wordline	negative for standby	[137][142]
WL driver V_{DD}	lower in standby	[131]
well biasing	change with mode	[137][141]
bitline V_{DD}	lower for standby	[144]

graphically symmetrical to reduce the impact of lithographical mismatch on delay at lower V_{DD}, similar to [150]. Although DVS can provide power reduction for active memories, the more common approach to voltage scaling is to implement it primarily for idle SRAM blocks.

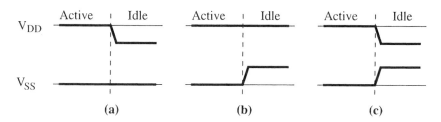

Fig. 7.22. General approaches using voltage scaling to lower idle power in SRAM. Lowering V_{DD} (a), raising V_{SS} (b), or both (c).

Reducing the voltage in an idle memory array lowers the standby power. Figure 7.22 shows three general methods for implementing standby voltage scaling. The power supply itself is reduced in Figure 7.22(a) (e.g. [132][136][137] [133]), the ground voltage is increased in Figure 7.22(b) (e.g. [143][144][145][146] [131][147]), and both rails are scaled in Figure 7.22(c) (e.g. [138]). For the case where V_{DD} is lowered for idle cells, the minimum voltage for retaining bistability was theorized in [1] and modeled for SRAM in [136]. Implementations of SRAM using lower V_{DD} in standby are available [137] along with software policies to determine when to enter the lower leakage mode [132].

One issue for deeply voltage scaled SRAM is Soft Error Rate (SER). Soft errors occur when an alpha particle or cosmic ray strikes a memory node and disrupts it such that it loses its value. Since the susceptibility of a memory node to a soft error is proportional to the amount of charge stored there, there is a minimum amount of capacitance that should be present in a cell to keep SER acceptably low [151]. Since bitcell storage capacitance decreases with scaling, and voltage scaling further reduces the stored charge, SER is a concern for sub-threshold memory. Fortunately, there are methods for taking care of soft errors. Studies of soft errors have shown that multi-cell errors from a single strike only occur in a limited number of cells along a wordline (2 to 3) [152]. Thus, physically interspersing bits from different words can prevent multi-errors from occurring in a single word [152]. The additional application of error correcting codes using parity bits can fix any such errors that occur. A chip that implements these techniques reports excellent success in reducing SER [144].

A common alternative to lowering V_{DD} for reducing standby leakage is to increase the virtual ground node, V_{SS}, as in Figure 7.22(b). By explicitly raising V_{SS} or allowing it to float to higher voltages, transistor V_{DS} reduces and RBB further reduces leakage by increasing device V_Ts. In [131] and [144], V_{SS} is raised explicitly to 0.3V and 0.5V, respectively. In [146], the virtual ground node is allowed to float, and its voltage is limited by a diode-connected transistor.

Scaling the supply rail voltages is often supplemented by other leakage reducing approaches for standby. For example, negatively biasing the wordlines reduces leakage into the bitcells through the access transistors [137][142]. The negative wordlines are combined with V_{DD} lowering and with N-well biasing to match pMOS and nMOS leakage currents in [137]. Further precautions such as nMOS pullups on the bitlines lighten the stress on the negatively driven access transistor gates. The delay associated with charging V_{DD} is estimated to be only 5% of the total read delay [137].

Well-biasing is also used specifically for leakage reduction in standby [153][141]. In [141], a triple-well process allows RBB during standby and Forward Body Bias (FBB) during active operation. The various voltages required for this approach are generated off chip, and thick oxide devices are required in some places to withstand larger than normal gate voltages.

Other memories play more tricks with voltage supplies. In [139][140], the V_{DD} to the bitcells always is boosted relative to the periphery's V_{DD} by 100mV, and the chip works to a periphery voltage of 0.4V. The higher V_{DD} at the cross-coupled inverters improves Read SNM and reduces read delay by strengthening the drive transistors relative to the access transistors. The same effect degrades the ability to write, so the pMOS transistors in the bitcell use higher V_T to allow robust write operation.

A large SRAM in [131] uses three supply voltages (0.3V, 0.8V, and 1.2V) to implement different operational modes as illustrated in Figure 7.23. During deep standby, V_{SS} increases to 0.3V, the wordline V_{DD} drops to 0.8V using

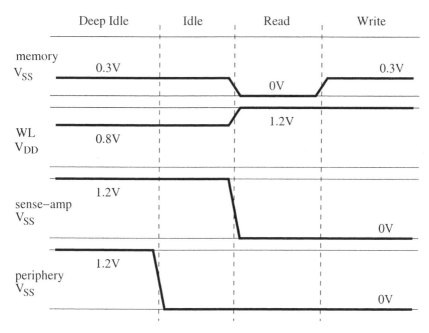

Fig. 7.23. Voltage scaling strategies for different operating modes in [131].

a diode-connected FET, and the sense amps and other peripherals are fully power gated by pulling their ground nodes to 1.2V. V_{SS} also is increased to an entire row during write access.

Clearly, previous efforts have explored many options for voltage scaling. However, none have yet pushed voltage scaling into the sub-threshold region during active operation.

7.2.2 6-Transistor SRAM Bitcell in Sub-threshold

Traditional 6T SRAMs face many challenges in DSM technologies, and low V_{DD} operation exacerbates the problems. This section describes key obstacles to sub-threshold SRAM operation for a 65nm process. Predictions in [154] suggest that process variations will limit standard 90nm SRAMs to around 0.7V operation for two primary reasons: degradation of SNM and reduced write margin. Variations in the bitcell transistors caused by phenomena such as global process variation, random doping mismatch, and temperature changes degrade the SNM. The impact of random local variation increases for DSM devices because of the smaller transistor channel area. In the sub-threshold region, variations in the threshold voltage impact delay and current exponentially. Previous work has measured the minimum voltage for retaining SRAM state during idle mode at several hundred millivolts for a 90nm memory [136].

Interestingly, the sensitivity of the SNM to threshold voltage mismatch actually decreases in the sub-threshold region [133], but the lower V_{DD} decreases the absolute value of SNM. Likewise, write access into the bitcell becomes less certain at lower supply voltages. Since standard write operation depends on a carefully balanced ratio of currents, processing variation makes this ratio difficult to maintain as V_{DD} decreases, leading to errors during write access.

These practical problems associated with low voltage operation for SRAMs limit the traditional 6T bitcell and architecture to higher voltage, above-threshold operation. Reports in the literature of 65nm SRAMs confirm this voltage barrier. A 65nm SRAM built in a dynamic-double-gate SOI (D2G-SOI) process functions to 0.7V in [155]. The authors analyze their design for bulk CMOS and report that it cannot operate below 1.0V [155]. A bulk CMOS 65nm SRAM also reports its minimum operating voltage as 0.7V [145].

Our results confirm that SNM degradation and inability to write are the two most significant obstacles to sub-threshold SRAM functionality in 65nm. We examine the SNM problem in greater detail in Section 7.2.5. This section examines the critical problems with write and read operation for a 65nm 6T SRAM in sub-threshold.

7.2.3 Write Operation

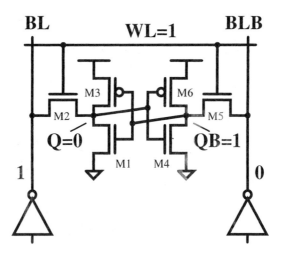

Fig. 7.24. Schematic showing conceptual write of '1' into node 'Q' in the 6T bitcell. Data shown for Q and QB must be overwritten.

Figure 7.24 shows a 6T bitcell at the onset of a write operation. The write drivers are applying the new bit values to the bitlines, but the cell still holds

7.2 Sub-threshold SRAM 125

the old values. In order for Q and QB to convert from their old values ('0' and '1') to their new values ('1' and '0'), the write drivers must overpower the feedback inside the cell. Since the nMOS access transistors, M_2 and M_5, are poor at driving a '1', the key to proper write operation lies with writing the new '0' correctly. This requirement presents a well-known sizing problem for above-threshold design. Specifically, there is a ratioed fight between the pMOS inside the cell that holds a '1' and the series combination of the nMOS access transistor and the write driver [125]. In Figure 7.24, the write driver on BLB and M_5 must overpower M_6 to write a '0' to node QB.

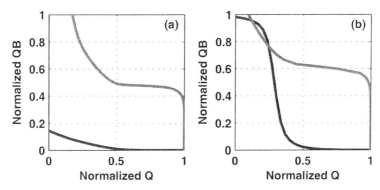

Fig. 7.25. Example VTCs for write access showing write 'SNM'. A negative write SNM indicates a successful write to the mono-stable point (a). If the write SNM remains positive, then the write fails because the state of the bitcell is not changed - example shown in (b).

When the write operation succeeds, the bitcell becomes mono-stable, forcing the internal voltages to the correct values. Figure 7.25(a) shows a butterfly curve of a bitcell displaying correct write operation. This is the same type of plot used to demonstrate SNM for the bitcell. During a successful write, there are no lobes on the butterfly curve. Using the SNM terminology, we can say that the 'write SNM' is negative. If the VTC and inverse VTC curves on the plot shift by an amount equal to the write SNM, then the cell will regain bi-stability. Figure 7.25(b) gives a different example of a positive write SNM that corresponds to a failure to write the bitcell. In this particular example, the static characteristics of the butterfly curve show that the write driver and access transistor cannot sufficiently overpower the bitcell's feedback to write Q='1'. The key to achieving a successful write in the traditional fashion, then, is to ensure that the access transistor and write driver win the fight with the pMOS pull-up inside the bitcell.

In the 65nm process for which we are designing, the pMOS mobility makes it weaker than an iso-sized nMOS at nominal V_{DD}, but the pMOS current in sub-threshold is larger than an iso-sized nMOS. This makes write functionality

126   7 Sub-threshold Memories

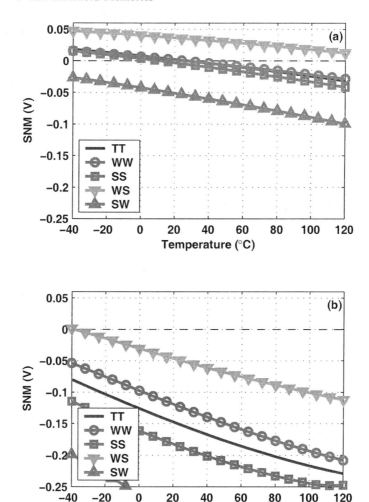

Fig. 7.26. SNM for write access versus temperature and process corner (TT, WW, SS, WS, and SW) at $V_{DD} = 0.3V$ (a) and $V_{DD} = 0.6V$ (b). Negative SNM indicates successful write.

more challenging. Figure 7.26 shows the write margin of a 6T bitcell versus temperature and process corner. Again, negative SNM corresponds to correct write functionality. At $V_{DD} = 300mV$ in Figure 7.26(a), the standard method of writing fails for large regions of process corner and temperature. The general trend showing an improvement of write operation (i.e. more negative write margin) at higher temperature occurs because the pMOS transistors weaken relative to nMOS as temperature rises. Thus, the access transistors become more capable of overcoming the pMOS that holds a '1' inside the bitcell. As supply voltage increases, the write margin improves. Figure 7.26(b) shows the write margin at 0.6V. This voltage is well above the V_T of both types of transistor, and the pMOS has weakened relative to the nMOS because the mobility starts to dominate the differences in V_T. Even at 0.6V, the write margin is barely negative for the worst-case corner, and this plot does not account for local V_T variation. For these reasons, $V_{DD}=0.6$V is the best case voltage for which we can expect traditional write operations to work for a sub-threshold memory in this 65nm process.

7.2.4 Read Operation

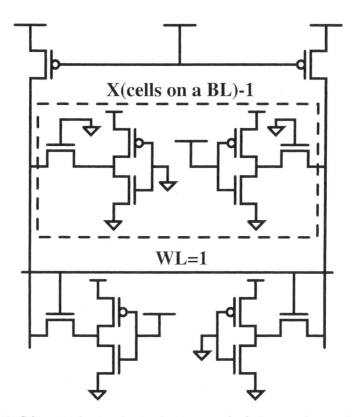

Fig. 7.27. Schematic showing the simulation set-up for finding steady-state behavior for the worst-case scenario for bitline leakage. For X cells on a bitline, $X-1$ cells hold the complement of the value in the accessed cell, maximizing leakage in opposition of the desired read.

128   7 Sub-threshold Memories

As described in Section 7.1, bitline leakage is a significant problem for DSM SRAM, just as it is for register files. Despite various techniques to minimize leakage into unaccessed cells, technology scaling leads to progressively shorter bitlines that have been segmented to reduce bitline leakage [134]. This section examines the impact of bitline leakage on read operation in the sub-threshold region for a 65nm technology.

Figure 7.27 shows the simulation set-up for examining the steady-state behavior for the worst-case bitline leakage. After the read access reaches steady state, the *on*-current of the access transistors for the addressed bitcell drives the bitlines. The leakage current of the remaining $X - 1$ bitcells opposes the accessed cell. The DC voltage on the bitlines shows the steady-state impact of bitline leakage for X cells on a bitline.

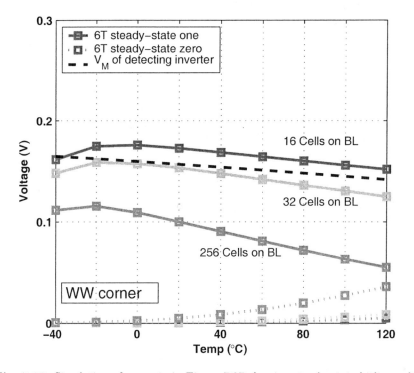

Fig. 7.28. Simulation of scenario in Figure 7.27 showing steady-state bitline voltages. Bitline leakage severely limits the number of cells that can share a bitline.

Figure 7.28 plots the results of a simulation using the set-up from Figure 7.27 for the 6T bitcell in 65nm CMOS at $V_{DD} = 300$mV. The bitline that should be '1' droops significantly because of bitline leakage into the other cells. This figure shows the WW corner which shows the worst-case, but the other corners do not give significant improvement. The droop gets worse at higher temperature because the pMOS transistors weaken relative to nMOS as temperature increases, strengthening the bitline leakage. The plot shows the switching threshold, V_M, of an inverter that can serve as a simple sense amplifier to detect the full-swing output of this bitcell. Clearly, a 6T SRAM using this inverter for sensing is limited to 16 bitcells on a bitline at best. Even more complicated sense amplifiers face the challenge of operating on a small bitline differential that develops slowly because of leakage.

The other, more fundamental problem for 6T bitcells in sub-threshold is degraded Read SNM. The next section deals with SNM in greater detail.

7.2.5 Static Noise Margin in sub-threshold

This section evaluates the Static Noise Margin (SNM) of 6T SRAM bitcells operating in sub-threshold. In [156], we analyze the dependence of SNM during both hold and read modes on supply voltage, temperature, transistor sizes, local transistor mismatch due to random doping variation, and global process variation in a commercial 65nm technology. In this work, we focus on the impact of process variations.

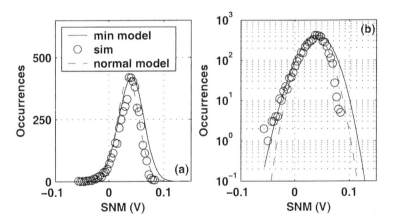

Fig. 7.29. Histogram of Read SNM Monte-Carlo simulation (circles) with normal PDF (dash) and min PDF model ([156]) (solid) over-laid. The semilog plot (b) shows that the PDF based on the model in [156] matches the worst-case tail quite well. (© 2005 IEEE)

Since the SNM during a read access is worse than during hold, the worst-case tail of the Read SNM distribution limits the yield of large memories. Figure 7.29 shows an example of this distribution for the case of random local variations in the six transistors of the bitcell. A model from [156] gives a good estimate of the worst-case tail of this distribution.

The impact of this local V_T variation on Read SNM imposes a serious limitation on SRAM V_{DD} scaling. Increasing the sizes of the transistors in the bitcell and/or raising V_{DD} is necessary to achieve the desired statistical yield. This problem suggests that changing the bitcell to eliminate the Read SNM problem would allow lower V_{DD} operation and thus decrease both total energy and standby power.

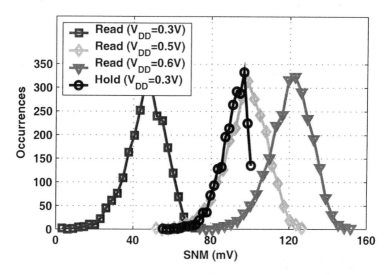

Fig. 7.30. Distribution of Hold SNM at 300mV compared with Read SNM distributions at different voltages. Read SNM at 500mV has the same mean, but it has a larger standard deviation.(© 2006 IEEE)

Figure 7.30 shows the distribution of the Read and Hold SNMs for a 6T bitcell at a 300mV supply voltage. The mean Read SNM is only slightly above half of the mean Hold SNM, but, even worse, the deviation of the Read SNM is larger than for the Hold SNM. For a multiple megabit memory, numerous cells will have Read SNM less than zero based on this statistical analysis. From this figure, the mean of the Read SNM at 500mV roughly equals the mean of the Hold SNM at 300mV. However, it is unclear from this plot how the Hold SNM and Read SNM compare at the worst-case tails.

Figure 7.31 shows the CDF functions derived from the distributions using the model in [156]. These CDF curves show how Hold SNM compares to Read SNM. For 6σ probability, the Hold SNM for a given V_{DD} roughly equals the Read SNM for twice that V_{DD} in the range of interest for us. This means that a memory that avoids the Read SNM problem can operate at roughly half of the V_{DD} of a 6T memory with the same 6σ bitcell stability. The fact that by avoiding the limitation imposed by Read SNM we can operate at lower

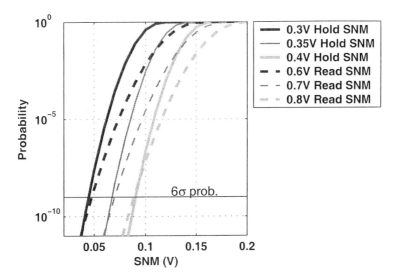

Fig. 7.31. CDF of SNM distributions showing that avoiding the Read SNM allows a reduction in V_{DD} by ~ 0.5 for the same 6σ stability.

voltages with the same cell stability is a key observation used in the design of a sub-threshold bitcell.

7.2.6 A Sub-threshold Bit-cell Design

The previous sections in this chapter point to several advantages of an SRAM bitcell that can operate into the sub-threshold region. Previously published works have scaled SRAM V_{DD} into the sub-threshold region during idle, but no SRAM actually operates in this region. The 0.18μm memory in [103] provides one exception. This memory operates down to 180mV, deep into the sub-threshold region. In structure, it resembles a register file more nearly than a standard 6T SRAM, as we described in Section 7.1.3. The bitcell itself looks like a latch with a tristate driver for writing into the cell and a tristate inverter replacing a standard inverter in the cross-coupled inverter pair so that it can cut off the feedback during write. For read, the outputs of a pair of cells are multiplexed together, and these outputs are successively multiplexed until the addressed word is selected. These muxes correspond to a bitline shared by only two bitcells. Accounting for the multiplexors required to read, the equivalent bitcell size is 18 transistors. Despite its large size, this sub-threshold SRAM benefits from the robustness of full-swing read and static, non-ratioed write operations.

Taking this previous implementation [103] as a datapoint in the set of possible bitcells, we can set up a range of bitcell options. At one end of the

132   7 Sub-threshold Memories

range, the 6T bitcell cannot operate below 600-700mV in 65nm. At the other end of the spectrum, an 18T bitcell will function robustly in sub-threshold since it looks and functions very much like combinational logic. Along the range in between these two options are many possible bitcell designs that address the obstacles to sub-threshold operation by increasing the number of transistors relative to the 6T cell. The bitcell that this section describes was selected from among many others because it represents the best trade-off of functionality and area. In other words, it is the smallest bitcell from those examined that provides robust sub-threshold functionality.

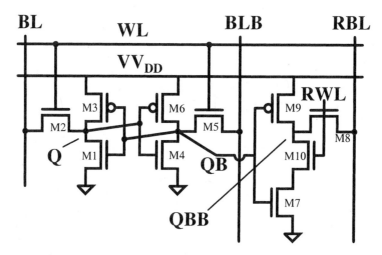

Fig. 7.32. Schematic of the 10T sub-threshold bitcell.(© 2006 IEEE)

Figure 7.32 shows the schematic of a 10T bitcell that addresses these problems and provides sub-threshold functionality. Transistors M_1 through M_6 are identical to a 6T bitcell except that the source of M_3 and M_6 tie to a virtual supply voltage rail, VV_{DD}. Write access to the bitcell occurs through the write access transistors, M_2 and M_5, from the write bitlines, BL and BLB. Transistors M_7 through M_{10} implement a buffer used for reading. Read access is single-ended and occurs on a separate bitline, RBL, which is precharged prior to read access. The wordline for read also is distinct from the write wordline. One key advantage to separating the read and write wordlines and bitlines is that a memory using this bitcell can have distinct read and write ports. Since a 6T bitcell does not have this feature, the 10T bitcell is in some ways more fairly compared to an 8T dual-port bitcell (6T bitcell with two pairs of access transistors and bitlines). The remainder of this section describes the operation of the 10T sub-threshold bitcell in detail.

7.2 Sub-threshold SRAM 133

Enabling Sub-threshold Read

The 10T bitcell in Figure 7.32 uses transistors M_7-M_{10} to remove the problem of Read SNM by buffering the stored data during a read access. When the read wordline (RWL) goes high, the pre-charged read bitline (RBL) causes a voltage divider across M_7, M_8, and M_{10}, but this increase in voltage at QBB does not impact the stored data at Q and QB. Thus, the worst-case SNM for this bitcell is the Hold SNM related to M_1-M_6, which is the same as the 6T Hold SNM for same-sized M_1-M_6. Eliminating the Read SNM problem allows this bitcell to operate at half of the V_{DD} of a 6T cell while retaining the same 6σ stability. A different approach for eliminating the Read SNM in [157] uses a 7T cell to prevent the higher voltage at the internal node from propagating to the other back-to-back inverter. This approach works well for strong-inversion operation, but it requires the bitcell to hold its data dynamically during read accesses. This approach will not work in sub-threshold because the dynamic data is susceptible to leaking away during the long access times.

The extra FETs in the 10T bitcell increase area by $\sim$ 66% (based on layout) and also consume leakage power relative to a 6T bitcell. It is interesting to note that a 9T bitcell, identical to the bitcell in Figure 7.32 but without M_{10}, would eliminate the Read SNM problem while using less area than the 10T cell. M_{10} is valuable to the bitcell because it reduces leakage current and it allows more bitcells to share a bitline.

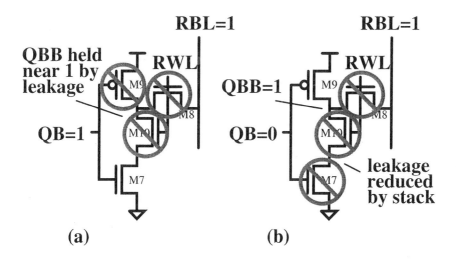

Fig. 7.33. Schematic of read buffer from 10T bitcell for both data values. In both cases, leakage is reduced to the bitline and through the inverter relative to the case where M_{10} is excluded.

134   7 Sub-threshold Memories

Figure 7.33 shows the read buffer from the 10T bitcell for Q=0 (a) and Q=1 (b). When Q=0 and QB=1 (Figure 7.33(a)), M_{10} adds an off device in series with the leakage path through M_8 and the path through M_9, decreasing the leakage through those transistors. Furthermore, since the pMOS in this 65nm technology generally has higher leakage than the nMOS, the leakage in M_9 tends to hold node QBB near V_{DD} (see Figure 7.34), further limiting the leakage through M_8 to the bitline by making its V_{GS} negative. Even if QBB floats above 0 by only a small amount, the negative V_{GS} in M_8 reduces bitline leakage exponentially. When Q=1 and QB=0 (Figure 7.33(b)), M_{10} creates a stack of *off* nMOS transistors, reducing leakage through M_7 by the stack effect. Since node QBB is held solidly at V_{DD}, M_8 has $V_{DS} = 0$, so bitline leakage is negligible. In both cases, M_{10} reduces the leakage relative to the 9T case.

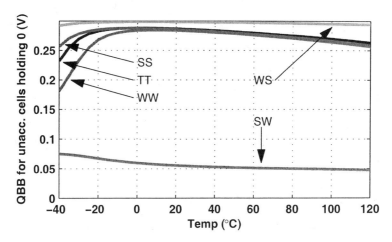

Fig. 7.34. Simulation of voltage at node QBB in unaccessed 10T bitcells versus temperature and process corner. Strong pMOS leakage holds QBB near V_{DD} except at the SW corner. Even at SW, QBB is higher than it is for the 6T cell, lowering bitline leakage.

Figure 7.35 shows the relative leakage of the different bitcells under consideration. At 0.3V and nominal conditions, the 9T bitcell has 50% leakage overhead relative to the 6T bitcell. The 10T bitcell reduces this overhead to 16%. It is important to recall that while the 6T bitcell can hold data at this low voltage, it cannot function properly for either read or write accesses. Since a 6T bitcell at 600mV has the same 6σ stability as a 10T bitcell at 300mV, this overhead in leakage current is more than compensated by decreasing V_{DD} by 300mV relative to the 6T bitcell. In simulation, the 10T bitcell at 300mV

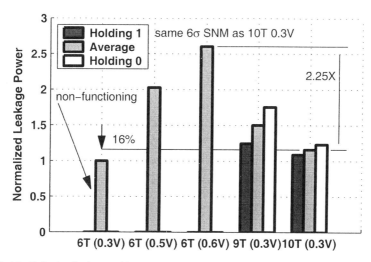

Fig. 7.35. Relative leakage of bitcells. The 10T bitcell imposes 16% overhead leakage at 300mV, but the 6T bitcell cannot function at that voltage (it can only hold data). The 10T cell saves 2.25X leakage power relative to the 6T at 0.6V.(© 2006 IEEE)

consumes 2.25X less leakage power than the 6T bitcell at 0.6V (1.75X less relative to 0.5V).

The reduction in sub-threshold leakage through M_8 reduces the impact of leakage from unaccessed cells and gives the additional advantage of allowing more cells on a bitline during read. As described in Section 7.2.4, bitline leakage creates real problems for SRAMs in terms of leakage power and functionality during a read access. Leakage from the bitline into the unaccessed bitcells causes undesirable voltage changes on the bitlines. Specifically, the bitline that should remain at its precharged value of V_{DD} will droop. For differential sensing, this droop creates an effective voltage offset that the accessed cell must overcome before activating the sense amplifier, which results in longer read access times. For single ended read access like that used with the 10T cell, the steady-state voltage values for a '1' and '0' become more difficult to distinguish.

Figure 7.36 shows the impact of bitline leakage on the steady-state voltages while reading a '1' (solid lines) or '0' (dotted lines). For the same number of cells on a BL, the 10T bitcell (circles) shows larger bitline separation than the 6T (or 9T) bitcells (squares). This figure suggests that 'sensing' with an inverter (whose switching threshold, V_M, is shown) should work well from 0^oC to 100^oC even with 256 cells on a bitline for the 10T cell. In contrast, as previously discussed in Section 7.2.4, the 6T cell (or 9T bitcell) would allow at most 16 bitcells on a bitline. The bitline that should be '1' stays very close to V_{DD} at high temperatures and then begins to droop at lower temperatures. This droop occurs because M_{10} inside the unaccessed 10T bitcells is so success-

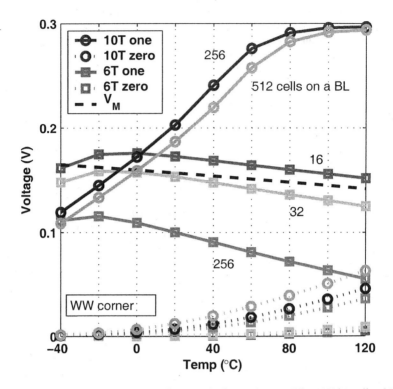

Fig. 7.36. Simulation showing steady-state bitline voltages. The 10T bitcell exhibits much better steady-state bitline separation than the 6T cell. The WW corner is shown at 300mV.

ful at reducing sub-threshold current through the access transistors that the sub-threshold current actually drops below the gate leakage (which is fairly constant with temperature). At higher temperatures, the leakage through the pMOS precharge device on the bitline exceeds the gate leakage into the unaccessed cells and holds the bitline close to V_{DD}. As temperature decreases, the sub-threshold leakage through the precharge transistor drops faster than the sum of gate leakage into the unaccessed cells until the accessed cell current must hold the bitline voltage high. At this point, the bitline voltage has to droop in order for the accessed cell to supply the current that flows into the other cells. If gate leakage was lower (perhaps in the case of high-K dielectrics), then sub-threshold leakage into the unaccessed cells is reduced sufficiently such that the bitline will stay very close to V_{DD}.

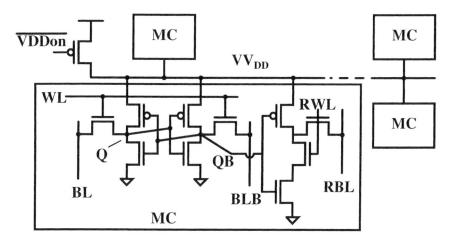

Fig. 7.37. Schematic of write architecture for a single row using a floating power supply (VV_{DD}). The row is 'folded' in layout so that its cells share N-wells, and the entire row is written at once. (© 2006 IEEE)

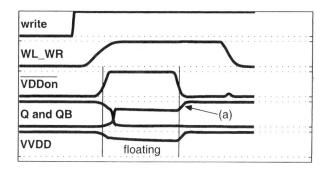

Fig. 7.38. Timing diagram for write operation. When $\overline{V_{DDon}}$ goes low while WL_{WR} remains asserted, the cell's feedback restores full voltage levels for the new values of Q and QB (point (a)).(© 2006 IEEE)

Enabling Sub-threshold Write

Write functionality offers the second primary obstacle to sub-threshold SRAM. In this 65nm technology, a 6T bitcell cannot write in the traditional fashion below around 0.6V, as we described in Section 7.2.3. The primary reason for write failure was the inability of the write driver and nMOS access transistor to win the ratioed fight against the pMOS inside the bitcell and to write a '0'. Previous work that use a virtual supply rail that floats during a write access [142][131][131][155][133]. Although those works applied this approach primarily for increasing speed, the method itself addresses the problem that sub-threshold bitcells face.

Since the pMOS devices in the 10T bitcell are the problem, our approach uses a virtual power supply rail rather than a virtual ground rail. Figure 7.37 shows the simple schematic for a single row using this approach. A single power-supply-gating header switch connects node VV_{DD} to the true power rail. When the bitcell holds its data or during read accesses, $\overline{V_{DDon}} = 0$ so that $VV_{DD} = V_{DD}$. During a write access, the virtual rail floats.

Figure 7.38 shows the timing associated with a write access using this scheme. First, the 'write' signal goes high to indicate that a write access will occur, and the bitlines (BL and BLB in Figure 7.37) are driven with the new data. Next, the decoders drive a global wordline (not shown) which eventually causes the local write wordline (WL_{WR}) to go high. Triggered by the local wordline, the $\overline{V_{DDon}}$ signal goes high, allowing node VV_{DD} to float. As the write access transistors discharge the virtual rail, its voltage droops, and Q and QB change to their new values. While VV_{DD} continues to float, denoted by the 'floating' label on the timing diagram, the logical '1' inside the cell tracks its drooping voltage value. When $\overline{V_{DDon}}$ goes low again while the *local wordline remains high*, it reconnects the virtual rail to the full supply. The feedback inside the bitcell then holds the Q and QB nodes at their correct logical values and amplifies the '1' to full V_{DD}. This occurs at point (a) in Figure 7.38.

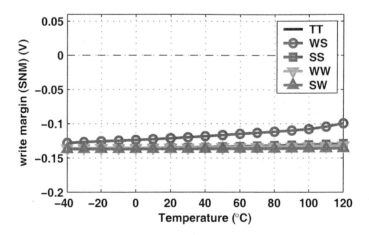

Fig. 7.39. Write margin (write SNM) versus temperature at 0.3V for 10T bitcell with floating VV_{DD} supply. Negative margin for all corners, signifying successful write operation.

Some previous works implement a floating rail in the column-wise direction. The risk of the column-wise approach is that any droop that occurs on

VV_{DD} during a write operation will impact other unaccessed bitcells that are holding their data. For sub-threshold operation, the lack of voltage headroom increases the risk of losing data in those cells by decreasing their Hold SNM. For this reason, we implement the virtual rail along a row of the memory, as Figure 7.37 shows. For the implementation on the test chip, a conceptual row is folded as shown in the figure so that its bitcells can share N-wells, and the entire row is written at once.

The plot in Figure 7.39 shows the write margin for the virtual V_{DD} approach across temperature and process corner at $V_{DD} = 300\text{mV}$. The write margin remains negative across all of these ranges, indicating a successful write. The worst-case write margin occurs at the WS corner and high temperature.

Since the 10T bitcell shows the ability to solve both the read and write problems for sub-threshold operation, we chose it as the bitcell for a test chip in 65nm bulk CMOS.

7.2.7 65nm Sub-threshold SRAM Test Chip

This section describes the test chip that uses the 10T bitcell to enable sub-threshold operation.

Test Chip Architecture

A 256kb 65nm bulk CMOS test chip uses the 10T bitcell and the architecture shown in Figure 7.40. The memory is divided into eight 32-kb blocks. Each block contains an array of 256 rows and 128 columns of 10T bitcells. A single 128-bit Data Input/Output (DIO) bus serves all eight blocks. In this initial instantiation of the sub-threshold memory, only one read or write can occur per cycle. As mentioned previously, however, the 10T bitcell can accommodate both a read and write access to the same block in a single cycle. Such a dual-port instantiation of the memory would require a second DIO bus and additional peripheral logic. The decoder in this memory uses the top three address bits to determine the block and generates a block select signal (BKsel) to enable certain local features within the selected block. The remaining eight address bits select the correct row inside the block. The decoder decodes these eight bits and asserts a global wordline. The global wordline then asserts a local wordline inside the selected block. The local wordline then combines with the local write signal to assert either WL_{RD} or WL_{WR}. For a write access, local logic turns off $M_P\langle r \rangle$ to the accessed row as described in Section 7.2.6. The write drivers consist simply of inverters with transmission gates. Although unnecessary for functionality in this design, the transmission gates turn off when the memory is not writing to minimize leakage on the write bitlines (BL and BLB). The power supply to the WL drivers is routed separately to allow a boosted WL voltage. This technique improves the access speed and increases the robustness to local variations. The read bitline (RBL) is precharged prior

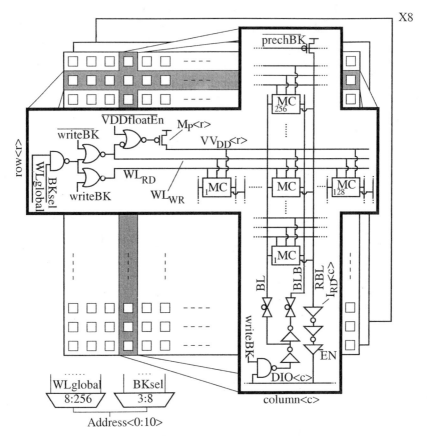

Fig. 7.40. Architecture diagram of the 256kb memory on the test chip using 10T sub-threshold bitcells.(© 2006 IEEE)

to read access, and its steady-state value is 'sensed' using a simple inverter, $I_{RD}\langle c \rangle$, as discussed in Section 7.2.6. Tristate buffers prevent the output of the blocks from driving the DIO bus at the incorrect times. Column and row redundancy is a ubiquitous technique in commercial memories used to improve yield. For our analysis of the SRAM, we assume the availability of one redundant row and column per block.

The primary goals for this test chip were to test the functionality of the 10T bitcell in sub-threshold and to explore the limitations of the design. For this reason, the peripheral circuits were designed to be as simple as possible. All of the peripherals use static CMOS logic for simplicity and for functional robustness in sub-threshold. The large block size was intentionally aggressive in order to expose limitations in the bitcell and architecture. Integrating

256 bitcells on the bitline (as opposed to 16 for 6T) pushes the envelope for functionality.

The layout of the memory had to meet logic design rules. For this reason, even the reference 6T bitcell layout is much larger than commercial 6T layouts for which the design rules are relaxed. The 10T bitcell layout added almost exactly the expected 66% area overhead relative to our reference 6T design. The area penalty for the entire array may not be so large, however. Since the 10T bitcell allows 256 bitcells on a bitline, fewer copies of the column-wise peripherals are necessary than for the 6T cell array, which can only accommodate 16 cells per bitline pair. As an example, the 8T bitcell in [158] has 40% larger area than its 6T counterpart. However, by equalizing bitline leakage [158], the bitcell allows 256 cells per bitline rather than 16. The total cache using the 8T bitcell in a 100nm technology ends up being smaller than its 6T counterpart by 6% [158]. This example suggests that the total array area penalty for the 10T cell is much less than 66%, since it gives a similar advantage in bitline integration.

Table 7.2. 6T and 10T architecture comparison.

line	6T		10T	
	number	transistors	number	transistors
write WL	1	256	1	256
write BL	2	256	2	256
read WL	1	256	1	256
read BL	2	256	1	256

As we described in Section 7.2.6, we chose for this implementation to switch the N-wells along a row along with VV_{DD}. This approach made it easier to follow the design rules related to distance between well taps and avoided the need to route an additional V_{DD} rail. To make this approach work, each row is folded such that a pair of 64-bit physical rows sharing N-wells and a VV_{DD} rail makes up one conceptual 128-bit row (c.f. Figure 7.37). This folding increases the length of bitlines by roughly 2X and decreases the length of wordlines by roughly $\frac{1}{2}$X. Notice that this is not fundamentally necessary for the write approach to work. The N-wells of two separate rows can be shared and the VV_{DD} for each row routed separately.

The impact of the 10T approach on the number of wordlines and bitlines used during memory accesses is beneficial. Table 7.2 shows the comparison for a 6T memory that also has 256 rows and 128 columns (this could not actually function in sub-threshold or even above threshold because of bitline leakage). The 10T approach uses one less bitline, and the transistor load on any given wordline or bitline is the same. As we mentioned previously, separate wordlines and bitlines for write and read accesses allow simultaneous write and read accesses to the memory.

142   7 Sub-threshold Memories

Figure 7.41 shows a layout shot and die photograph of the test chip. The die size is 1.89mm by 1.12mm, and the chip is pin-limited. The 256kb array and a 32kb block are highlighted for reference. Metal fill in all of the metal layers obscures the features in the interior of the die photograph.

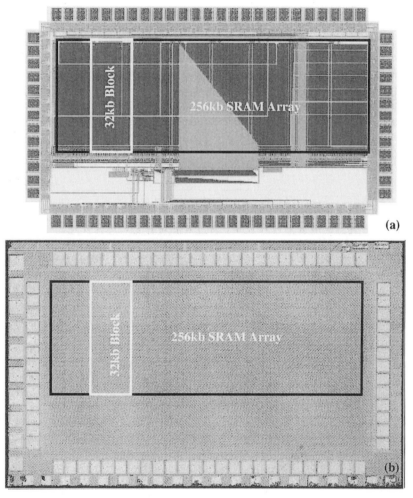

Fig. 7.41. Annotated layout (a) and die photograph (b) of the 256kb sub-threshold SRAM in 65nm. Die size is 1.89mm by 1.12mm.(© 2006 IEEE)

7.2 Sub-threshold SRAM   143

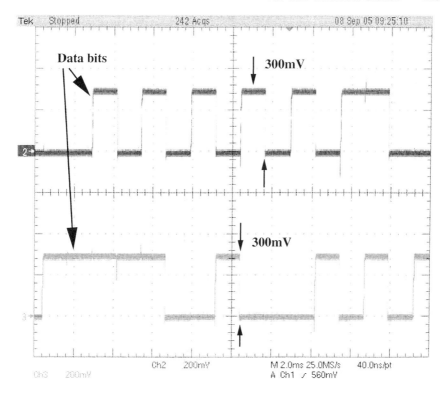

Fig. 7.42. Oscilloscope waveform showing correct functionality at $V_{DD} = 300$mV. At this low voltage, a small fraction of bits have errors.(© 2006 IEEE)

Measurements

Measurements of the SRAM test chip confirm that it is functional over a range of voltages from 1.2V down into the sub-threshold region. With the assumption of one redundant row and column per block, the memory operates correctly to below 400mV. Read operation works without error to 320mV and write operation works without error to 380mV at 27°C. We continued to push the supply voltage to even lower values to examine the limits of the implementation. At the low supply voltage of 300mV, the memory continues to function, but it does exhibit bit errors in ∼ 1% of its bits that result from sensitivities in the architecture to local device variation. Figure 7.42 shows an oscilloscope plot of two data bits output from the memory during read operations at 300mV.

The test chip successfully demonstrates a functional sub-threshold memory that overcomes the problems it was designed to face. First, the bitcell removes the Read SNM problem. Measurements have confirmed that the memory experiences zero destructive read errors at 300mV. Simulations show that a 6T memory would experience a high rate of destructive read errors at 300mV due to degraded Read SNM. Secondly, whereas a 6T memory would fail to write below about 600mV, this memory writes correctly at 350mV at 85°C. Thirdly, a 6T memory would experience problems reading with only 16 bitcells on a bitline. Measurements show that the 10T memory reads correctly even with 256 bitcells on the bitline down to 320mV. Finally, the memory shows good Hold SNM performance. The first bits observed to fail to hold their data occur at $V_{DD} < 250$mV.

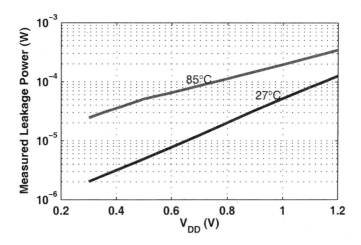

Fig. 7.43. Measured leakage power from the memory test chip.

Figure 7.43 shows the measured leakage power of the test chip at two different temperatures. As expected, voltage scaling provides significant reduction in leakage power. Figure 7.44 shows the relative savings in leakage power from V_{DD} scaling. The plot is normalized to operation at 0.6V for comparison to a theoretical minimum-voltage 6T memory. At 27°C, the 10T memory saves 2.5X and 3.8X in leakage power by scaling from 0.6V to 0.4V and 0.3V, respectively. Leakage power decreases by over 60X when V_{DD} scales from 1.2V to 0.3V. Voltage supply scaling also gives the expected savings in active energy. Figure 7.45 shows the energy per read access versus supply voltage based on the measured switched capacitance of the memory.

In summary, sub-threshold SRAM provides the dual advantages of minimizing total memory energy consumption and of providing compatibility with minimum-energy sub-threshold logic. Traditional 6T SRAM cannot function in sub-threshold because it fails to write below ∼ 600mV and because the Read SNM degrades badly at low supply voltage. Furthermore, bitline leakage in 6T SRAMs limits the number of bitcells on a bitline to around 16. A 10T bitcell solves these problems and provides functionality into the sub-threshold

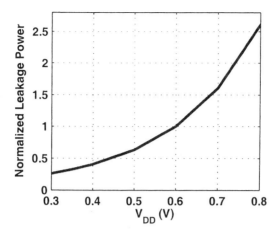

Fig. 7.44. Relative leakage power savings at 27°C achieved by V_{DD} scaling.(© 2006 IEEE)

region. The bitcell solves the write problem by using a floating supply voltage that allows the write drivers to overcome the cell feedback. A read buffer prevents read accesses from affecting the stored data and thus removes the Read SNM problem. The read buffer uses a low-leakage design that allows many more cells to use the same bitline relative to the 6T bitcell.

A 256kb 65nm bulk CMOS test chip uses the 10T bitcell and demonstrates sub-threshold operation. Measurements show that the bitcell fundamentally

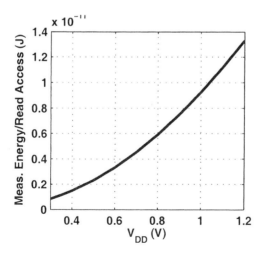

Fig. 7.45. Measured active energy per read access.

solves the Read SNM problem, overcomes the write problem, and relaxes the bitline integration limitation. With one redundant row and column per block and a boosted wordline, the memory functions without error to below 400mV. At 400mV, it consumes 3.28μW and works up to 475kHz. No bit errors for holding data occur in the SRAM until V_{DD} scales below 250mV.

8
Analog Circuits in Weak Inversion

by Eric A. Vittoz

8.1 Introduction

As explained in Chapter 5, the characteristics of a transistor in weak inversion are very different from those in strong inversion. Weak inversion has some special properties that can be exploited, in particular for designing low-power and/or low-voltage analog circuits. But weak inversion also has some drawbacks, the most important being the poor current matching (see Section 5.8.1), a maximum noise content of the drain current (see Section 5.6.4), and of course the low speed. Therefore, and except for very low supply voltages (below 0.5 V), the best performance is obtained by combining transistors biased in weak and in strong inversion.

It must be remembered that the amount of inversion is controlled by the inversion coefficient IC defined in Section 5.3.4. In principle, any value of IC can be obtained for any value of saturation current by adjusting the specific current I_{spec} (5.32) through W/L. The only limits are the width W of the transistor (and the associated drain-substrate leakage) for I_{spec} very large, and its length L (and the associated leakage *underneath the channel*) for I_{spec} very small.

This chapter will explore where and how the various favorable features of weak inversion can be exploited in designing analog circuits.

148   8 Analog Circuits in Weak Inversion

8.2 Minimum Saturation Voltage

8.2.1 Current Mirrors

As illustrated by Figure 5.8, weak inversion provides the minimum possible saturation voltage. The most immediate application is that of low-voltage current mirrors. The output current becomes independent of the output voltage as soon as this voltage is larger than 4 to $6U_T$. It is thus possible, in principle, to build voltage amplifiers for a supply voltage as low as 8 to $12U_T$, provided the threshold voltage is sufficiently low.

The price to pay for this low voltage drop across the mirror is a low speed, a maximum spectral density for both channel noise and interface noise, and the worst matching of currents.

The only known possibility to drastically improve the matching and virtually eliminate the flicker noise, while keeping the low minimum output voltage, is to build the mirror with bipolar transistors. Those might be MOS transistors operated in the lateral bipolar mode, that are available in any CMOS process [159].

8.2.2 Cascode Mirrors

The residual output conductance of a mirror can be drastically reduced by cascoding it by means of a common-gate transistor, as illustrated in Figure 8.1.

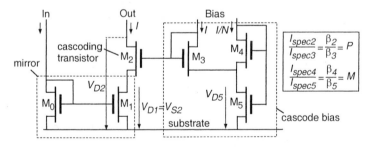

Fig. 8.1. Cascode mirror in weak or moderate inversion.

The common-gate transistor M_2 must be biased so as to ensure the saturation of the mirror transistor M_1. This is the purpose of the sub-circuit M_3-M_4-M_5 (cascode bias), itself biased by currents equal or proportional to the output current I of the mirror.

If transistor M_1 is in *weak inversion*, it is saturated for $V_{D1} = 4$ to $6U_T$ *proportional to the absolute temperature* T (PTAT). If all the other transistors are also in weak inversion, then equation (5.40) applied to each transistor gives [110, 106]

8.2 Minimum Saturation Voltage

$$V_{D5} = U_T \ln\left[1 + (N+1)M\right], \tag{8.1}$$

$$V_{D1} = V_{D5} + U_T \ln P = U_T \ln\left[P(1 + (N+1)M)\right]. \tag{8.2}$$

The product MN should not be too large, to prevent the drain junction leakage of a wide transistor M_4 from dominating a very small bias current I/N.

Now, even for a large V_{D1}, the output conductance of M_1 cannot be lower than G_{ds}, the residual conductance due to channel shortening given by (5.50). We can therefore assume that saturation is obtained when the residual drain transconductance due to the remaining reverse current I_R is smaller than G_{ds}. Hence, according to (5.44):

$$G_{md1} = \frac{I_{R1}}{U_T} < G_{ds1} = \frac{I_{F1}}{V_{M1}} \quad \text{or} \quad \frac{I_{R1}}{I_{F1}} < \frac{U_T}{V_{M1}}. \tag{8.3}$$

By using the expression (5.39) of I_F and I_R, this gives

$$V_{DS1} > U_T \ln \frac{V_{M1}}{U_T}, \tag{8.4}$$

or by using (8.2) where $V_{S1} = 0$:

$$P(1 + M(N+1)) > V_{M1}/U_T. \tag{8.5}$$

For $V_{M1}/U_T < 200$, this condition can be fulfilled by choosing $N = 2$ and $P = M = 8$, which gives $V_{D1} = 5.3 U_T$.

The circuit of Figure 8.1 can also be used for *moderate inversion*, provided a different approach is used for sizing the devices. This approach is based on the fact that the saturation of M_1 is ensured if $I_{R1} \ll I_{F1}$, *independently* of the inversion coefficient IC [160, 109]. For this purpose, we choose $M_5 \equiv M_1$ (two identical transistors) and $M_3 \equiv M_2$ ($P = 1$). Furthermore, $N \gg 1$ (which is possible in spite of the drain junction leakage, since the transistors are not in deep weak inversion).

Now M_2 and M_3 are both saturated with the same current density. Since they also have the same gate voltage they have the same source voltage, thus $V_{D1} = V_{D5}$. Transistors M_1 and M_5 are not fully saturated, but since $N \gg 1$, they have the same drain voltage and the same current density. Therefore:

$$I_{F1}/I_{R1} = I_{F5}/I_{R5}, \tag{8.6}$$

and the two transistors are in the same state of saturation.

The forward current I_{F4} of M_4 and the reverse current I_{R5} of M_5 are controlled by the same voltage V_{D5}. Since M_4 is M-times wider than M_5, then

$$M I_{R5} = I_{F4} = I/N \quad \text{whereas} \quad I_{F5} = I \tag{8.7}$$

since $I_{R5} \ll I_{F5}$. Thus

$$\frac{I_{F1}}{I_{R1}} = \frac{I_{F5}}{I_{R5}} = MN. \tag{8.8}$$

This ratio could be further increased by choosing $P > 1$, but the effect would then depend on the inversion coefficient [160].

If enough voltage is available, it is interesting to use this circuit with M_1 (and therefore M_0 and M_5) in strong inversion to improve current matching, but M_2 (and therefore M_3) in weak inversion to minimize V_{DS2} required for saturation.

Recalculating the condition (8.3) for $G_{md} < G_{ds}$ with the expression of transconductance (5.43) valid at all levels of current gives

$$\frac{I_{R1}}{I_{F1}} < \frac{U_T}{V_{M1}}\left(1 + \frac{U_T}{V_{M1}}IC_1\right), \tag{8.9}$$

which only departs from (8.3) for a very large inversion coefficient IC_1.

A simpler bias circuit can be used if both M_1 and M_2 are in strong inversion [161].

8.2.3 Low-Voltage Amplifiers

Thanks to the minimum saturation voltage, MOS transistors in weak inversion can provide some voltage gain A_v even at a very low supply voltage V_B. The maximum gain is achieved by the CMOS inverter as illustrated in Figure 8.2.

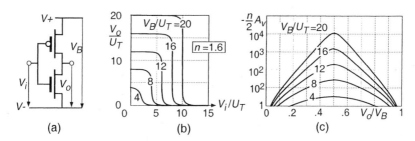

Fig. 8.2. CMOS inverter-amplifier: (a) circuit and definitions; (b) transfer characteristics $V_o(V_i)$; (c) gain as a function of the output voltage V_o.

Assuming that the two complementary transistors have the same values of n and $V_{T0} + nV_S$ in the equation (5.40) of the drain current, the unloaded transfer characteristics can be obtained by equating the current of the two transistors [53]. This gives

$$v_i = \frac{v_B}{2} + \frac{n}{2}\ln\frac{1-e^{v_o-v_B}}{1-e^{-v_o}}, \tag{8.10}$$

where v_B, v_i and v_o are the normalized supply, input and output voltages according to

$$\frac{V_B}{v_B} = \frac{V_i}{v_i} = \frac{V_o}{v_o} = U_T. \tag{8.11}$$

These characteristics are plotted in Figure 8.2(b) for several values of v_B.

The voltage gain A_v can then be obtained by differentiating (8.10), which yields

$$A_v = \frac{dv_o}{dv_i} = -\frac{2}{n} \cdot \frac{e^{v_o - v_B} + e^{-v_o} - e^{-v_B} - 1}{2e^{-v_B} - e^{v_o - v_B} - e^{-v_o}}. \tag{8.12}$$

It is plotted in Figure 8.2(c) for several values of v_B. As can be seen, a voltage gain larger than 100 can be obtained with a supply voltage of $12U_T \cong 300\text{mV}$. For this symmetrical circuit, the gain reaches a maximum value for $V_i = V_o = V_B/2$ given by

$$|A_v|_{max} = (e^{v_B/2} - 1)/n. \tag{8.13}$$

The residual output conductance in saturation (G_{ds} due to channel shortening or G_{mdsat} due to DIBL) has been omitted in this calculation and would put another limit on very high values of A_v.

By combining such a CMOS inverter with an adequate biasing circuit [40], it can be used as an analog amplifier.

In spite of the relatively low speed associated with weak inversion, transistors in RF front ends of integrated receivers can be biased close to weak inversion for low-voltage operation thanks to the availability of short channel lengths in submicron processes [162].

For digital applications, a supply voltage of $4U_T \cong 100\text{mV}$ is sufficient to provide the necessary nonlinear transfer characteristics, as shown by Figure 8.2. Of course, the maximum current should be adjusted to obtain the required speed. This can be done, in principle, by choosing a very low value of V_{T0} and by controlling the current by the source voltage V_S [163, 53]. This should be done separately for P- and N-channel transistors in order to achieve the expected symmetrical characteristics.

8.3 Maximum Transconductance-to-Current Ratio

8.3.1 Differential Pair

As a consequence of the maximum value of G_m/I_D obtained in weak inversion, the difference of gate voltages required to compensate the mismatch of two transistors is minimum (see Figure 5.14). Moreover, the spectral density of input referred noise voltage is also minimum for a given value of drain current (see Section 5.6.4). These features can be exploited to minimize the offset and the noise of differential pairs, with the additional advantage of minimum saturation voltage.

It should be noticed that, as explained at the end of Section 5.8.1, the source voltage should be as small as possible in order eliminate the effect of Δn (mismatch of slope factors). This means that, for minimizing the input voltage offset, the two transistors should be put in a separate well connected to their sources, as illustrated in dotted line in Figure 8.3(a). In single-well processes, this is unfortunately only possible for one type of transistors (P-channel for N-well process).

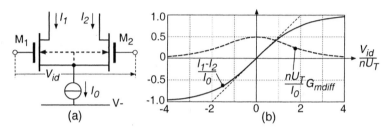

Fig. 8.3. Differential pair; (a) circuit; (b) transfer characteristics and transconductance.

The transfer characteristics of the saturated differential pair in weak inversion can be calculated by using equation (5.40) of the drain current, giving

$$I_1 = \frac{I_0}{1 + \exp\frac{-V_{id}}{nU_T}} \quad \text{and} \quad I_2 = \frac{I_0}{1 + \exp\frac{+V_{id}}{nU_T}}, \tag{8.14}$$

or for the difference of output currents

$$I_1 - I_2 = I_0 \tanh\frac{V_{id}}{2nU_T}, \tag{8.15}$$

which is represented in Figure 8.3(b). As can be seen, a major drawback of a differential pair in weak inversion is its limited input range of linearity. This is best illustrated by the differential transconductance

$$G_{mdiff} = \frac{d(I_1 - I_2)}{dV_{id}} = \frac{I_0}{2nU_T} \cdot \left(\cosh\frac{V_{id}}{2nU_T}\right)^{-2} \tag{8.16}$$

plotted in the same figure.

The linear range can be extended by using the circuit of Figure 8.4(a) [164], which can be analyzed by means of the concept of pseudo-resistors introduced in Section 5.5, as shown by Figure 8.4(b). The saturated transistors M_1 and M_2 correspond to grounded resistors R_1 and R_2 in the resistor prototype. The linearization transistors M_{1a} and M_{2a} correspond to two resistors of K-times larger values. The difference of output currents can now be calculated by using the resistor prototype, giving

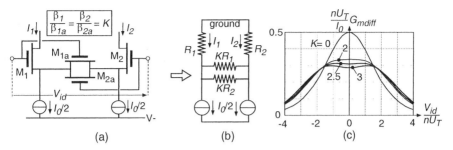

Fig. 8.4. Linearized differential pair; (a) circuit; (b) resistor prototype; (c) differential transconductance.

$$\frac{I_1 - I_2}{I_0} = \frac{R_2 - R_1}{R_2 + R_1 + K \cdot R_2 R_1/(R_2 + R_1)} = \frac{e^{2x} - 1}{e^{2x} + (K+2)e^x + 1}, \quad (8.17)$$

where the last part has been obtained by replacing R_2/R_1 by e^x, with $x = V_{id}/(nU_T)$, according to the definition (5.54) of pseudo-resistors in weak inversion. The transconductance is obtained by differentiation of (8.17):

$$G_{mdiff} = \frac{d(I_1 - I_2)}{dV_{id}} = \frac{I_0}{nU_T} \cdot \frac{(2+K)e^{2x} + 4e^x + (2+K)}{(e^{2x} + (2+K)e^x + 1)^2} e^x, \quad (8.18)$$

which reduces to (8.16) for $K = 0$. This result is plotted in Figure 8.4(c) for several values of K. As can be seen, the best linearity is obtained for $K \cong 2.5$, at the price of a 40% reduction of transconductance.

Another method for increasing the linear range is the "multi-tanh" technique, which was developed for bipolar transistors [165]. This approach uses the sum of the currents of differential pairs having symmetrical controlled offset voltages, as illustrated in Figure 8.5 for 2 pairs. The specific current

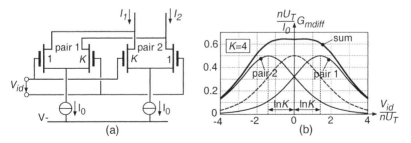

Fig. 8.5. Multi-tanh linearization: (a) circuit; (b) transconductance (single symmetrical circuit in dotted line for comparison).

of the two transistors of each pair are in a ratio K, which produces in weak inversion, according to (5.40), an input offset voltage

$$\Delta V_{id} = nU_T \ln K. \tag{8.19}$$

As shown in Figure 8.5(b), $K = 4$ (which can easily be implemented by 4 unit transistors) is an optimum value. The range of linearity can be further extended by choosing $K = 13$ and adding the output currents of a normal differential pair biased at $0.75I_0$ [165].

8.3.2 Single-Stage Operational Transconductance Amplifiers (OTA)

Another consequence of the maximum value of G_m/I_D obtained in weak inversion is a maximum value V_M/U_T of the intrinsic voltage gain, as was shown in Figure 5.10. If the channel is not too short, this voltage gain can be as high as 60 dB per stage and can be boosted to close to 120 dB by cascoding the current sources. Hence, Operational Transconductance Amplifiers (OTA's) can be implemented in a single cascoded stage, which eliminates the need for compensation.

In weak inversion, the unity gain bandwidth G_m/C_L is maximum for a given load capacitance C_L and a given current, therefore the settling time is minimum. But this is only true for small signals. Indeed, if a large voltage step is applied to the input, the differential pair saturates, hence the settling time is set by the slow slew-rate associated with the small bias current.

To circumvent this problem, the bias current I_0 of the differential pair must be increased whenever its input voltage V_{id} is large. One way to do this is to provide an increment of I_0 proportional to the difference of output currents [42] as illustrated in Figure 8.6(a), where

$$I_0 = I_b + A|I_1 - I_2| \quad \text{with} \quad I_0 = I_1 + I_2. \tag{8.20}$$

If the pair operates in weak inversion, the combination of (8.20) with (8.14)

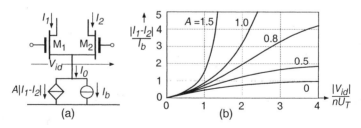

Fig. 8.6. Adaptive biasing of a differential pair: (a) circuit; (b) transfer characteristics for several values of feedback factor A.

gives

$$|I_1 - I_2| = I_b \cdot \frac{\exp\left|\frac{V_{id}}{nU_T}\right| - 1}{(A+1) - (A-1)\exp\left|\frac{V_{id}}{nU_T}\right|}, \tag{8.21}$$

8.3 Maximum Transconductance-to-Current Ratio

which is plotted in Figure 8.6(b) for several values of A. With $A = 0$, it is the basic differential pair that saturates at $|I_1 - I_2| = I_b$. This saturation current increases for $0 < A < 1$. For $A > 1$, $|I_1 - I_2|$ tends to infinity (i.e. it is no longer limited by the bias current I_b) for a critical value of input voltage given by

$$|V_{idcrit}| = nU_T \cdot \ln \frac{A+1}{A-1}. \qquad (8.22)$$

For the particular case $A = 1$:

$$\min(I_1, I_2) = I_b/2, \qquad (8.23)$$

one of the two branch currents remains constant at $I_b/2$.

A special application of this adaptive biasing technique is the class AB voltage follower illustrated in Figure 8.7(a). Without the additional bias cur-

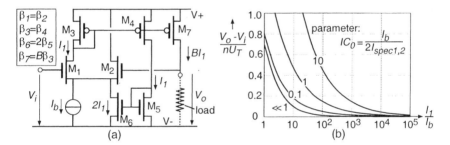

Fig. 8.7. Voltage follower for a resistive load: (a) circuit; (b) input-output offset.

rent $2I_1$ delivered by M_5-M_6, the circuit would be an elementary OTA with unity voltage feedback. It would only operate as a follower if the current delivered to the load were much lower than the maximum output current BI_b.

With the feedback through M_5-M_6, the total bias current of the differential pair M_1-M_2 is

$$I_0 = I_b + 2I_1 \text{ with } I_0 = I_1 + I_2, \text{ hence } I_2 = I_b + I_1. \qquad (8.24)$$

The difference between the input voltage and the voltage applied to the load can then be expressed by using (5.35):

$$\frac{V_i - V_o}{nU_T} = \frac{V_{P1} - V_{P2}}{U_T} = \sqrt{1 + 8IC_0 \cdot I_1/I_b} - \sqrt{1 + 8IC_0(1 + I_1/I_b)}$$

$$+ \ln \frac{\sqrt{1 + 8IC_0 \cdot I_1/I_b} - 1}{\sqrt{1 + 8IC_0(1 + I_1/I_b)} - 1} < 0, \qquad (8.25)$$

where $IC_0 = I_b/2I_{spec1,2}$ is the inversion coefficient of M_1 and M_2 without feedback. These characteristics are represented in Figure 8.7(b) for several values of IC_0. The output voltage is always *larger* than the input voltage, but the difference becomes very small for $I_1 \gg I_b$. For a given value of I_1/I_b, the difference is minimum if $IC_0 \ll 1$ (OTA without feedback in weak inversion).

A high current efficiency can be obtained by choosing $B \gg 1$ (ratio of mirror M_3-M_7).

As long as the differential pair remains in weak inversion, (8.25) reduces to

$$\frac{V_i - V_o}{nU_T} = -\ln\left(1 + \frac{I_b}{I_1}\right). \tag{8.26}$$

8.4 Exponential Characteristics

8.4.1 Voltage and Current Reference

The exponential dependency of the drain current on V_S/U_T makes it possible to extract a voltage proportional to U_T as shown in Figure 8.8. The basic

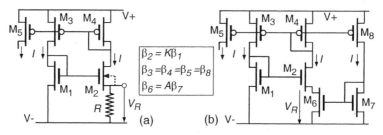

Fig. 8.8. Voltage and current reference: (a) basic circuit; (b) resistor-less current reference.

circuit [31] shown in part (a) of the figure contains a 1-to-K N-channel current mirror M_1-M_2, with the source of M_2 degenerated by a resistor R. A 1-to-1 P-channel current mirror M_3-M_4 (or any equivalent circuit) imposes the same current in the two branches. Therefore, a source voltage V_R builds-up across resistor R to compensate the ratio K of the N-channel mirror. If this mirror is in weak inversion, then

$$V_R = RI = U_T \ln K. \tag{8.27}$$

This voltage should be sufficiently larger than the threshold mismatch of M_1-M_2. In practice, it cannot be made much larger than $4U_T$, which corresponds to $K = 55$. It can be used as a PTAT voltage reference, or as a compensation voltage in a band gap voltage references [32].

A reference current I can be extracted by the additional mirror transistor M_5. Thanks to the small value of V_R a small current can be obtained with a reasonably low value of R.

If the transistor M_2 is in a separate well connected to its source, as shown by the dotted line, then $V_{S2} = V_{S1} = 0$. The factor K is then compensated by a difference of gate voltages, and U_T is multiplied by n in (8.27).

Figure 8.8(b) shows a variant of the basic circuit in which the resistor is replaced by transistor M_6 [166]. This transistor is the output transistor of a current mirror M_7-M_6 of ratio 1 to A, that operates in *strong inversion*, with the reference current itself as its input. If $A \gg 1$, then $I_{F6} = AI_{F7} = AI \gg I$. Hence, far from being saturated, M_6 is biased close to $V_D = V_S = 0$ where it behaves like a resistor of value $R = 1/G_{ms6}$ given by (5.45):

$$1/R = G_{ms6} = \sqrt{2n\beta_6 AI}. \tag{8.28}$$

By introducing this value in (8.27) we obtain, after arranging the result

$$I = 2n\beta_6 U_T^2 \cdot A(\ln K)^2 = I_{spec6} \cdot A(\ln K)^2. \tag{8.29}$$

This current is obtained without using any resistor, and is proportional to the specific current of transistor M_6. Therefore, as noticed at the end of Section 5.7, it becomes almost independent of the temperature if the mobility is proportional to $T^{-\alpha}$ with $\alpha \cong 2$.

In practice, (8.28) is an acceptable approximation for $A > 5$.

It should be mentioned that the loop M_6-M_2-M_4-M_8-M_7 implements a positive feedback. However, the gain of this loop can be shown to be 1/2 at equilibrium.

8.4.2 Amplitude Regulator

The exponential characteristics of transistors in weak inversion are exploited in the amplitude regulator depicted in Figure 8.9(a) [31]. The sinusoidal signal of amplitude V produced by the oscillator enters the regulator through capacitor C_1, and transistor M_5 delivers the output current used to bias the oscillator. When no oscillation is present ($V = 0$), the circuit is reduced to the current

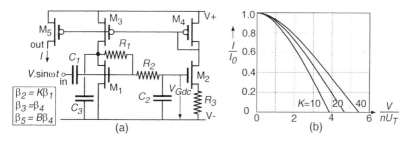

Fig. 8.9. Amplitude regulator for oscillators: (a) circuit; (b) transfer characteristics.

reference of Figure 8.8(a). According to (8.27), it delivers an output current

$$I = I_0 = BU_T \ln K / R_3, \tag{8.30}$$

which serves as the start-up current of the oscillator. As the oscillation voltage grows, it is superimposed on the DC component of gate voltages V_{Gdc} for M_1, but not for M_2, since it is blocked by the low-pass filter R_2-C_2. Because of the exponential function $I_D(V_G)$ of M_1, its average drain current should increase, which is not compatible with the 1-to-1 ratio imposed by the mirror M_4-M_3. Instead, V_{Gdc} decreases, resulting in a decrease of the output current I.

Assuming that all transistors remain saturated, that M_1 remains in weak inversion even during the peaks of its drain current, and that the residual oscillation amplitude at the gate of M_2 is much smaller than U_T (so that it has no nonlinear effect), the transfer characteristics are given by [31]

$$I = I_0 \cdot \left(1 - \frac{\ln I_{B0}(\frac{V}{nU_T})}{\ln K}\right), \tag{8.31}$$

where I_{B0} is the 0-order modified Bessel function. They are plotted in Figure 8.9(b) for several values of K.

The amplitude of oscillation will stabilize when the regulator delivers exactly the bias current I required to produce the amplitude V.

If the peak drain current of M_1 leaves weak inversion, the transfer characteristics will be modified and may eventually loose their monotonicity, which must absolutely be avoided to maintain stable oscillation. A semi-empirical condition to ensure monotonicity is

$$\beta_1 > \frac{2\beta_3/\beta_4}{nU_T R_3}. \tag{8.32}$$

The role of capacitor C_3 is to keep the drain voltage of M_1 sufficiently constant to avoid de-saturation during the positive peaks of current.

At low frequencies, high (non-critical) values may be needed for resistors R_1 and R_2. Very high values have been obtained by using lateral diodes in the polysilicon layer [167, 37]. Lower values can be obtained by means of transistors adequately biased [46].

This amplitude regulator has been applied extensively to quartz oscillators in watches [37, 46, 168, 169, 170], but it can be used in different type of sinusoidal oscillators as well [38].

8.4.3 Translinear Circuits

Discovered for bipolar transistors, the translinear principle [171] is an outstanding application of the exponential characteristics of MOS transistors in weak inversion. Consider the loops of saturated transistors illustrated in Figure 8.10. They include an even number of transistors, half of which have their gate to source "junction" in the clockwise (cw) direction, the other half in the counter-clockwise (ccw) direction. Hence, for the whole loop:

8.4 Exponential Characteristics

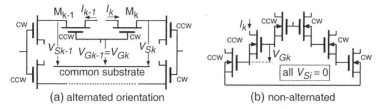

Fig. 8.10. Translinear loops: (a) alternated orientation of transistors in a common substrate; (b) non-alternated orientation: separate local substrates are necessary.

$$\sum_{cw}(V_{Gi} - V_{Si}) = \sum_{ccw}(V_{Gi} - V_{Si}). \tag{8.33}$$

If all transistors are in weak inversion, with negligible reverse current (saturated), then according to (5.41):

$$I_i = I_{D0i} \exp \frac{V_{Gi}/n_i - V_{si}}{U_T} \quad \text{or} \quad \frac{V_{Gi}}{n_i} - V_{Si} = U_T \ln \frac{I_i}{I_{D0i}}. \tag{8.34}$$

Now if cw and ccw transistors are *alternated* [172] as in Figure 8.10(a), then each gate voltage V_{Gi} is common to a pair cw-ccw of transistors. It appears therefore in both sides of equation (8.33), which can thus be rewritten as

$$\sum_{cw}(\frac{V_{Gi}}{n_i} - V_{Si}) = \sum_{ccw}(\frac{V_{Gi}}{n_i} - V_{Si}). \tag{8.35}$$

We can now introduce (8.34), divide by U_T (that is common to all transistors) and exponentiate both sides of the equation, which yields

$$\prod_{cw} \frac{I_i}{I_{D0i}} = \prod_{ccw} \frac{I_i}{I_{D0i}} \quad \text{or} \quad \frac{\prod_{cw} I_i}{\prod_{ccw} I_i} = \frac{\prod_{cw} I_{D0i}}{\prod_{ccw} I_{D0i}} = \lambda. \tag{8.36}$$

This result is independent of the temperature. If I_{D0} is the same for all transistors, then $\lambda = 1$. A circuit may include several loops sharing some transistors, each loop characterized by its value of λ. The mismatch of I_{D0i} is dominated by that of V_{T0i} and results in an error in the value of λ, with a standard deviation

$$\frac{\sigma(\Delta \lambda)}{\lambda} = \frac{1}{nU_T}\left[\sum \frac{1}{2}\sigma^2(\Delta V_{T0i})\right]^{1/2}, \tag{8.37}$$

where the factor 1/2 comes from the fact that $\sigma(\Delta V_{T0})$ is defined for a pair of transistors.

The current-mode multiplier/divider [173] shown in Figure 8.11(a) is a simple example of a single translinear loop with identical transistors ($\lambda = 1$) of alternated orientations.

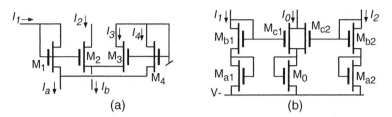

Fig. 8.11. Example of translinear circuits: (a) multiplier/divider; (b) vector length calculation.

The input currents are I_1, I_a and I_b whereas I_2 is the output. Using (8.36), we can write

$$I_1 I_3 = I_2 I_4 \quad \text{hence} \quad I_1(I_b - I_2) = I_2(I_a - I_1), \tag{8.38}$$

which gives after simplification

$$I_2 = I_1 I_b / I_a. \tag{8.39}$$

This result is valid as long as $I_1 < I_a$.

If cw and ccw transistors are *not* alternated, as in the loop example of Figure 8.10(b), then (8.33) cannot be rewritten as (8.35). The only way to make (8.34) compatible with (8.33) is to impose $V_{si} = 0$ by putting each transistor in a separate well connected to its source. Although each n_i is slightly dependent on the particular gate voltage, it can be approximated by a constant n multiplying U_T in (8.34).

In addition to requiring separate wells, non-alternated loops need a higher supply voltage, since they include stacks of at least two gate-to-source voltages.

An interesting example of a loop where transistors cannot be alternated is the circuit of Figure 8.11(b) that calculates the length of a vector in a N-dimensional space [174].

For this example with $N = 2$, the circuit contains two loops that share the transistor M_0. The corresponding current equations are

$$I_1^2 = I_{c1} I_0 \text{ and } I_2^2 = I_{c2} I_0, \text{ thus } I_0 = I_{c1} + I_{c2} = (I_1^2 + I_2^2)/I_0. \tag{8.40}$$

Hence, finally:

$$I_0 = \sqrt{I_1^2 + I_2^2}. \tag{8.41}$$

N loops are needed for N dimensions, each loop made of transistors M_{ai}, M_{bi} and M_{ci}, and all loops sharing M_0.

Although the basic translinear principle assumes that all the loop transistors are saturated, it is possible to include non-saturated transistors [175]. Consider M_k and M_{k-1} in the basic loop of Figure 8.10(a). They have the same gate voltage. Therefore, if they are connected in parallel, they represent the forward and reverse currents of a non-saturated transistor. This may help reducing the supply voltage needed by translinear circuits [176].

8.4.4 Log-Domain Filters [177, 178, 179, 180]

Consider the elementary circuit of Figure 8.12(a), where the transistor is assumed to be in weak inversion and saturated.

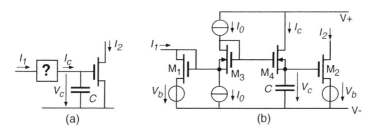

Fig. 8.12. Log-domain filters: (a) principle ; (b) direct implementation of the integrator (partial).

How should the input current I_1 be transformed into the capacitor current I_c in order to obtain a linear current-mode integrator, with

$$I_2 = \frac{1}{\tau} \int I_1 dt, \quad \text{or} \quad I_1 = \tau \frac{dI_2}{dt}, \tag{8.42}$$

in spite of the exponential relationship $I_2(V_c)$? According to (5.41):

$$I_2 = I_{D0} \exp \frac{V_c}{nU_T}, \quad \text{hence} \quad V_c = nU_T \ln \frac{I_2}{I_{D0}}. \tag{8.43}$$

The current in the capacitor can therefore be expressed as

$$I_c = C \frac{dV_c}{dt} = \frac{CnU_T}{I_2} \cdot \frac{dI_2}{dt} = \frac{CnU_T}{I_2} \cdot \frac{I_1}{\tau}, \tag{8.44}$$

where the last term is obtained by introducing (8.42). Thus, the circuit will behave as a linear integrator according to (8.42) if

$$I_c I_2 = I_0 I_1 \quad \text{with constant} \quad I_0 = CnU_T/\tau. \tag{8.45}$$

This relation could be implemented by the translinear loop illustrated in Figure 8.12(b) [181]. However, in this direct implementation the transistors are not alternated, and two of them must be put in a separate well connected to their source. Notice that it is anyway only a partial implementation, since no current is available to discharge the capacitor.

A variant [182] with alternated transistors is depicted in Figure 8.13. The current I_c flowing through M_3 is mirrored into C by M_5-M_6. The bias voltage V_b provides the necessary headroom for the controlled current source M_7.

Here again, the circuit shown in full line can only charge the capacitor C. One way to circumvent the problem is to add the current source I_d sketched

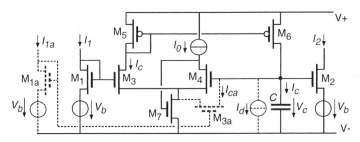

Fig. 8.13. Log-domain integrator implemented by an alternated translinear loop.

in dotted line. The current through M_3 is then $I_c + I_d$ and the loop equation becomes

$$I_1 I_0 = (I_c + I_d) I_2 \quad \text{or} \quad I_1 = \frac{I_c I_2}{I_0} + \frac{I_d}{I_0} I_2, \qquad (8.46)$$

which corresponds to the s-domain transfer function

$$\frac{I_2(s)}{I_1(s)} = \frac{1}{I_d/I_0 + s\tau}. \qquad (8.47)$$

It is thus a lossy integrator, that can be used to realize some specific filters [183]. The amount of damping could in principle be reduced by re-injecting an adequate proportion of I_2 at the input, but such a compensation is limited by the inaccuracy of current mirrors.

A better solution [182] is to implement a second loop formed of M_{1a}-M_{3a}-M_4-M_2 as also depicted in dotted line in Figure 8.13. The current through M_{3a} can now discharge C. The two input currents I_1 and I_{1a} are delivered by a signal conditioner that separates the positive and negative half-waves of the input signal.

8.5 Pseudo-Resistor

8.5.1 Analysis of Circuits

The concept of pseudo-resistor can facilitate the analysis of current-mode circuits by transforming them in a resistive prototype. In general, this possibility is limited to circuits made of transistors sharing the same gate voltage, like attenuators or special current mirrors [113, 109]. But, as was explained in Section 5.5, this limitation does not exist in weak inversion. If several transistors (in a common substrate) have the same gate voltage, the corresponding pseudo-resistances are proportional to their respective specific currents. For a difference ΔV_G of gate voltages, the ratio of pseudo-resistances is multiplied by $\exp\left(-\Delta V_G/nU_T\right)$ according to (5.54).

We have already seen an example of application in the linearized differential pair of Figure 8.4. Another example is illustrated by Figure 8.14. Consider

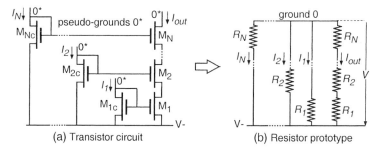

Fig. 8.14. Calculation of the harmonic mean of N currents.

the transistor circuit of Figure 8.14(a). It has N input currents I_1 to I_N and produces the output current I_{out}. All transistors are identical, some of them saturated, others not (M_1 to M_{N-1}). This circuit could be analyzed by introducing equation (5.41) for each transistor and by solving the resulting set of equations. Another possibility is to use the translinear principle, with N loops. The non-saturated transistors would have to be split in their forward and reverse components, as explained at the end of Section 8.4.3.

A much simpler approach uses the resistor prototype (or resistor equivalent) shown in in Figure 8.14(b). Since all transistors are identical, each pair M_i-M_{ic} corresponds to a pair of resistors of value R_i. As explained in Section 8.4.3, the saturated side of a transistor is a pseudo-ground (labeled 0*), that corresponds to a real ground (common node 0) is the resistor prototype (notice that, since we have N-channel transistors, all voltages are negative in the prototype). By inspection of this simple circuit, we obtain

$$I_{out} = \frac{V}{\sum_{i=1}^{N} R_i} = \frac{V}{\sum_{i=1}^{N}(V/I_i)} = \frac{1}{\sum_{i=1}^{N}(1/I_i)} = \frac{I_{hm}}{N}. \qquad (8.48)$$

The output current is proportional to the *harmonic mean* I_{hm} of the N input currents I_i. If the sum of these input currents is forced to be constant, I_{out} is maximum when all I_i are equal (coincidence or "bump" circuit [184]).

8.5.2 Emulation of Variable Resistive Networks

Thanks to the concept of pseudo-resistors, any network of variable linear resistors can be implemented by transistors operated in weak inversion, provided only currents are considered.

Each pseudo-resistor implemented by a transistor M can be controlled by a current by associating a control transistor M_c of same size, as illustrated in Figure 8.15. The pseudo-conductance $G^* = 1/R^*$ is then proportional to the control current I_c [112]. If the bias voltage V_b is the same for all pseudo-resistors of a circuit, then $G_i^*/G_j^* = I_{ci}/I_{cj}$ (same proportionality constant).

Both transistors must remain in weak inversion. Thus, for M≡M_c:

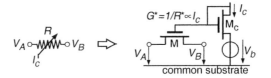

Fig. 8.15. Current control of a pseudo-resistor R^* corresponding to a real resistor R.

$$I_c \ll I_{spec} \quad \text{and} \quad V_A, V_B \geq V_b. \tag{8.49}$$

If needed, several pseudo-resistors may be controlled by the same control current through the same control transistor.

One of the most immediate applications of controlled resistors is the multistage variable linear attenuator depicted in Figure 8.16. The amount of at-

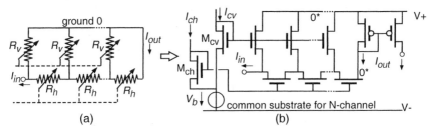

Fig. 8.16. Current-mode variable linear attenuator: (a) resistor circuit; (b) pseudo-resistor implementation.

tenuation depends on the number of stage and on the ratio $R_h/R_v = I_{cv}/I_{ch}$. Notice that the bias voltage V_b is needed to provide sufficient voltage headroom for the source of input current. The output current is extracted by means of a P-channel current mirror, which does not affect the current if the last horizontal transistor (hence all vertical transistors as well) are saturated (pseudo ground 0^*). If needed, the current through the vertical branch of each cell can also be extracted by a mirror.

The same circuit may be configured in a two-dimensional array to obtain a diffusion network. Some elementary spacial processing can then be carried-out on an image: low-pass spatial filtering is obtained by injecting the current of each pixel into the local node and by extracting it from the local vertical branch. Edge enhancement can then be obtained by subtracting each output from each input [50, 111]. A modification of the cell can provide local adaptation in order to eliminate gradients of illumination [185, 186].

An interesting novel application of pseudo-resistors in weak inversion is the on-line minimization of the energy spent by a multiprocessor system-on-chip to execute a set of related tasks [187].

8.5 Pseudo-Resistor

According to Maxwell's heat theorem [188], at steady state, any network of linear resistors driven by a constant current minimizes its power dissipation. The basic idea is to emulate the total energy E_{tot} required by the system to execute the whole set of tasks within a fixed duration D_{tot} by the power P_{Rtot} dissipated by a resistor network driven by a current I_{tot}.

Consider the example illustrated in Figure 8.17. Part (a) of the figure shows the task graph of 5 tasks T_1 to T_5 executed by 2 processing elements PE1 and PE2. The duration of the whole process is D_{tot}.

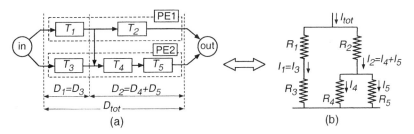

Fig. 8.17. Example of 5 tasks executed by 2 processing elements: (a) task graph; (b) corresponding network of resistors.

Now, each task T_i requires a given number N_i of cycles, and each cycle consumes an amount of energy that can be reduced if the supply voltage V_i is reduced. But the cycle time T_{ci} increases with V_i decreasing, and should therefore take the maximum value compatible with the available duration (no waiting time):

$$T_{ci} = D_i/N_i. \tag{8.50}$$

In this example, the task T_4 can only start after T_1 is ended, hence $D_1 = D_3$. The rest of the available duration D_{tot} should be entirely used for T_2 and for the two consecutive tasks T_4 and T_5. Hence $D_2 = D_4 + D_5$.

Figure 8.17(b) shows the corresponding network of resistors, in which each duration D_i of a task T_i corresponds to the current I_i in a resistor R_i. The total duration D_{tot} corresponds to the total current I_{tot} driving the circuit.

Now, the energy E_i per task and the power P_{Ri} per resistor are given respectively by

$$E_i = P_{Pi} D_i = \frac{P_{Pi}}{D_i} D_i^2 \iff P_{Ri} = R_i I_i^2 \tag{8.51}$$

where P_{Pi} is the (average) power consumption of the processor executing the task T_i. Hence, since the current I_i is proportional to the duration D_i

$$\frac{P_{Pi}}{D_i} \iff R_i \quad \text{thus} \quad G_i = \frac{1}{R_i} \propto \frac{I_i}{P_{Pi}} \tag{8.52}$$

Each resistor R_i is implemented as a pseudo-resistor R_i^*, so that its value can be adjusted proportionally to this ratio by means of a feedback loop that includes a calculation of P_{Pi}. This loop is illustrated in Figure 8.18. A key el-

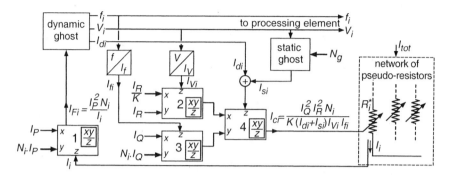

Fig. 8.18. Block diagram of a single control loop for energy minimization.

ement of the loop is a dynamic *ghost* circuit that mimics the maximum speed of the processing element built on the same chip and operated at the same voltage V_i. This ghost circuit is essentially a ring oscillator made of the same gates as the processing element. It is forced to oscillate at the required frequency $f_i = 1/T_{ci}$ by the input current $I_{Fi} \propto N_i/I_i$ in accordance with (8.50). It delivers the corresponding supply voltage V_i (which is the voltage needed to reach the frequency f_i) and the associated dynamic current consumption I_{di}. The voltage V_i and the frequency f_i are transmitted to the processing element. They are also converted to currents I_{Vi} and I_{fi} for further processing in the loop.

If necessary, a static ghost circuit is added to mimic the static current consumption of the processor, proportional to the total number of gates N_g (that may be different for different processors). This current is added to the dynamic current I_{di}.

Current-mode processing is carried out by multipliers/dividers (labeled xy/z). Each of them is implemented by the simple translinear loop of Figure 8.11(a) with some fixed bias currents I_P, I_Q and I_R.

The result is the current I_{ci} that controls the corresponding pseudo-resistor as illustrated in Figure 8.15. Hence

$$G_i = \frac{1}{R_i} \propto I_{ci} = \frac{I_Q^2 I_R^2 N_i}{K(I_{di}+I_{si})I_{Vi}I_{fi}} \propto \frac{I_i}{P_{Pi}} \tag{8.53}$$

as in (8.52). Notice that I_i (that becomes the input of the dynamic ghost through multiplier 1) cannot be used directly to produce the control current I_{ci}. It must be replaced by $N_i/I_{fi} \propto I_i$ that is the output of the dynamic ghost (response to the input).

The factor K introduced by multiplier 2 is proportional to the equivalent switching capacitance $P_P/(fV)$, that may be different for different processors.

9
System Examples

This chapter gives two examples of systems that operate in sub-threshold. First, a sub-threshold FFT is described. The FFT system employs the concept of energy-aware architectures. Energy-aware architectures contain hooks that allow the user to gracefully scale energy and quality depending on the system conditions. Exercising these power hooks causes changes in the workload and activity factor of the system, which in turn impacts its energy characteristics. Measurements of the FFT test chip show that, as activity factor varies, the minimum energy point also changes.

Second, a Local Voltage Dithering (LVD)-UDVS system extends the voltage range of traditional DVS systems. LVD improves on existing voltage dithering systems by taking advantage of faster changes in workload and by allowing each block to optimize based on its own workload. Additionally, measurements show that the time and energy overhead of LVD are small. UDVS also provides a practical method for extending DVS into the sub-threshold region. For many emerging energy-constrained applications, lowering energy consumption is the primary concern under most conditions. Thus, operating at the minimum energy point conserves energy at the cost of lower performance (frequency). This type of application works at the minimum energy point primarily and only jumps to higher performance voltages in rare cases. Chip measurements have shown the effectiveness of UDVS for this scenario.

9.1 A Sub-threshold FFT Processor

Sub-threshold operation is, as previous discussed, well-suited for wireless sensor nodes. The lifetime of a sensor node depends on the battery capacity and the ability of the node to compute and communicate in an energy-efficient fashion. Communication through an RF link expends a great deal more energy than computation, therefore an efficient sensor system performs sensor signal processing on the data and only transmits the resulting necessary information. This reduces the sensor bandwidth considerably and leads to longer sensor lifetimes.

To achieve minimal energy dissipation, a sub-threshold DSP is needed for sensors. A highly flexible node architecture contains both hardware accelerators and a programmable DSP [14]. Hardware accelerators can do signal processing algorithms extremely efficiently both in speed and energy dissipated. Sensor signal processing algorithms that are commonly used can be implemented in accelerators. The programmable DSP provides system control and implements functions that are not covered by the accelerators.

Examples of algorithms used in sensor nodes are beamforming, classification, direction sensing, etc. The Fast Fourier Transform (FFT) is an algorithm that is commonly used in sensor signal processing. This section shows details of a sub-threshold FFT implementation. The FFT uses various concepts described in Chapter 6 and Chapter 7. First, an energy-aware FFT architecture is described [189]. Energy-awareness provides power hooks into the architecture to allow the user to trade-off between energy and quality. This becomes important for sensors that must operate over a wide range of operating conditions. Next, the energy-performance contours for the FFT are analyzed. The energy-performance analysis describes the optimal operating point which occurs in the sub-threshold region at 400mV. Finally, chip measurements show functionality of the FFT down to 180mV and the optimal operating voltage as a function of varying the FFT bit precision and computation length.

9.1.1 The Fast Fourier Transform

The Fast Fourier Transform (FFT) is a widely used algorithm that appears in applications such as speech processing, signal detection, communications, and tracking. The FFT extracts the frequency and phase information from the sensor signals. Dedicated low-power FFT processors are able to sustain low-power requirements of various embedded applications [190].

Sensor data is considered "real-valued," which means that the imaginary part is zero. Traditionally, the FFT assumes complex input data. When an FFT is performed on real-valued data, the output is conjugate-symmetric. This means that, for an N-point FFT, the first N/2 points are unique, and the last N/2 points are symmetrically redundant. This symmetry is exploited by the real-valued FFT algorithm (RVFFT). The Real-Valued FFT (RVFFT) uses the symmetries inherent to computing the complex-valued FFT (CVFFT) on real-valued inputs to reduce overall computation. For example, a 1024-pt. RVFFT is efficiently performed by computing a 512-pt. CVFFT and then transforming the outputs back for 1024-pt. The computation effort of the RVFFT is approximately half that of the CVFFT.

A simple architecture for a RVFFT processor consists of a traditional CVFFT followed by backend processing. Figure 9.1 shows the conventional radix-2 butterfly architecture for the CVFFT using in-place computation. In-place computation occurs when a value is read out from the memory and is

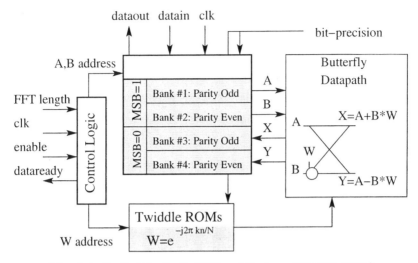

Fig. 9.1. Radix-2 butterfly FFT architecture. (© 2005 IEEE)

re-written to the same location in the next clock cycle. The advantage of in-place computation is that it requires the minimum sized buffer. For a vehicle-tracking sensor application, the maximum memory size was 1024-Words x 16-bit, which is the size of one frame of data. The FFT processor memory uses the register file design in Section 7.1. The memory was simulated and designed to operate to 100mV with a typical transistor model.

Additionally, a read-only memory (ROM) is needed for the storing twiddle factors (W). Twiddle factors are complex values used in the FFT to shift the phase of the input values.

9.1.2 Energy-Aware Architectures

In a sensor network, the environment is constantly changing. Energy-aware architectures are used to efficiently trade-off between energy and quality given current operating conditions. For example, a sensor with a full battery can provide very high quality sensor results and performance. When the battery energy is low, then the sensor can output lower quality results at a lower energy consumption and stretch out its battery lifetime.

The FFT architecture is designed with various power hooks that allow the architecture to gracefully scale FFT length and bit precision. On the same architecture, the sensor can perform a short 128-pt. FFT with 8-bit precision for a low-quality/low-energy result or perform a 1024-pt. FFT with 16-bit precision for a high-quality/high-energy result.

One of the hooks designed into the energy-aware FFT processor is variable bit-precision. The concept of variable bit-precision is showcased by the Baugh-Wooley (BW) multiplier in the butterfly datapath. A traditional fixed bit-precision BW design optimizes for the worst case scenario by building a single multiplier for the largest bitwidth. This design is non-optimal because, when lower precision multiplications are performed, the sign extension bits cause significant switching energy overhead. The proposed scalable BW multiplier design recognizes that the MSB quadrant contains a lower bit-precision multiplier. (The MSB quadrant includes those gates associated with the MSB inputs).

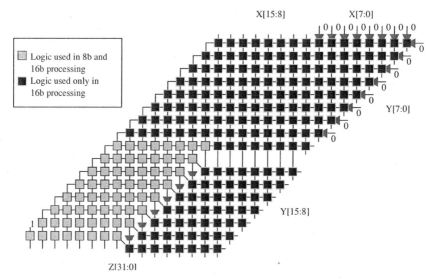

Fig. 9.2. 8b and 16b scalable Baugh-Wooley multiplier. (© 2005 IEEE)

To minimize switching in the LSB adders, the LSB inputs are gated, and only the MSB inputs are used to process data. Figure 9.2 demonstrates this technique for a BW multiplier that is scalable for 8-bit and 16-bit precisions. Similar bit-precision scalability was applied to the entire butterfly datapath, data memories and Twiddle ROMs. There is a 44% savings for 8-bit processing, but at the expense of 3% cost at 16-bit processing. 16-bit energy-aware processing has overhead due to the additional gates added to enable energy-awareness.

9.1.3 Minimum Energy Point Analysis

The FFT processor is designed to operate at the optimal operating point that minimizes energy dissipation. Analysis of the energy and performance of the FFT shows that the minimum energy point occurs at supply voltage levels below the threshold voltage. The energy and performance of the FFT were analyzed based on the theory discussed in Section 4.1. Figure 9.3 shows

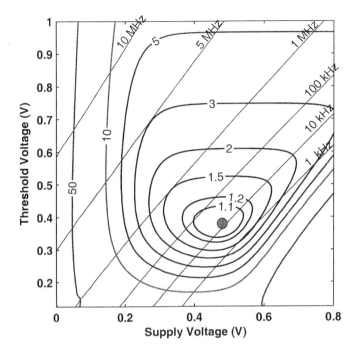

Fig. 9.3. Minimum energy point and constant energy and performance contours of the 16-b and 1024-pt. FFT. (© 2005 IEEE)

simulated energy contours of the 16-bit 1024-pt. FFT for a supply voltage range of 100mV-1V and threshold voltage range of 0V-800mV. The FFT was designed in a 0.18μm process. The figure shows the simulated average energy per FFT and performance across the entire supply and threshold voltage range using the switching and leakage models from Section 4.2.1.

The energy contours (circular) show that the minimum energy dissipation point occurs at V_{DD}=380mV and V_T=480mV. The performance contours show the frequency at the minimum energy point to be 13kHz.

The FFT processor is designed and fabricated in a standard 0.18μm bulk CMOS process with a fixed nominal threshold voltage of 450mV. Figure 9.4 shows an energy simulation at a fixed threshold voltage of 450mV. The predicted minimum energy point of the FFT processor occurs at 400mV. The FFT was designed and simulated at voltages much below 400mV to allow a thorough exploration of the space around the minimum energy point.

9.1.4 Measurements

The FFT processor was designed using a sub-threshold standard cell library, custom multiplier generators, and custom register file and ROM generators.

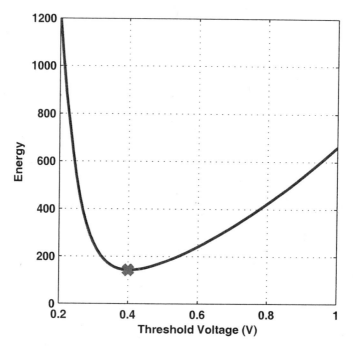

Fig. 9.4. Estimated minimum energy point for fixed V_T=450mV. (© 2005 IEEE)

The generators use specialized logic cells and ensure compact layout. The functional logic blocks in the datapath, control logic and memories were synthesized using the cell library.

The 0.18-μm CMOS FFT processor occupies 2.6 X 2.1 mm^2 and contains 627K transistors. It is fully functional at 128, 256, 512, and 1024 FFT lengths and for 8-b and 16-b precision, at voltage supply levels from 180 to 900mV with clock frequencies of 164Hz to 6MHz at these respective voltages. The power dissipated at 180mV is 90nW for 16-b 1024-pt. operation.

Figure 9.6 shows the measured energy consumption for 8-bit and 16-bit processing as a function of voltage. 8-bit processing has a lower activity factor and thus has lower switching energy. However, because the leakage energy is the same for both 8-b and 16-b processing, the minimum energy point increases. The minimum energy point for 16-bit occurs at 350mV and for 8-bit occurs at 400mV. The power dissipated at the 16-b optimum is 600nW at a clock frequency of 10 kHz, and the energy dissipated is 155nJ/FFT.

For an FFT sensor application benchmark, it proved to be 350X more energy efficient than a typical low-power microprocessor and 8X more energy efficient than a standard ASIC implementation.

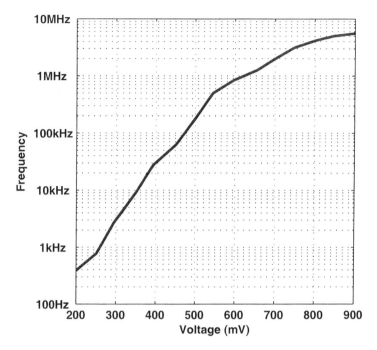

Fig. 9.5. Clock frequency as a function of voltage. (© 2005 IEEE)

9.2 Ultra-Dynamic Voltage Scaling

DVS has become a standard approach for reducing power when performance requirements vary. DVS systems lower the frequency and voltage together to reduce power when lower performance is allowed [191]. In the first true DVS implementation, a critical path replica is used in a feedback loop to adjust the supply voltage to the lowest value that allows the delay to match a given reference frequency [191]. DVS now appears in commercial processors such as, for example, the Intel XScale [192], IBM PowerPC [193], and the Transmeta Crusoe processor [194].

Voltage dithering was proposed as a low overhead implementation of DVS to provide near-optimum power savings using only a few discrete voltage and frequency pairs [68]. The savings are only achievable if the voltage and frequency can change on the same time scale as the altering workload. Previous implementations apply voltage dithering to entire chips and require many microseconds to change operating voltage [68][195]. This section describes a 90nm test chip that demonstrates a proposed concept of Local Voltage Dithering (LVD) and couples LVD with sub-threshold operation to achieve Ultra-Dynamic Voltage Scaling (UDVS) [196]. We provide measurements of

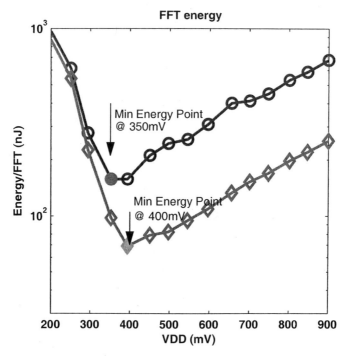

Fig. 9.6. Effect of activity factor on minimum energy point for 8-bit and 16-bit processing.

the effect of temperature on minimum energy operation for the 90nm test chip.

9.2.1 DVS and Local Voltage Dithering

Many signal processing systems process blocks of data that arrive at some regular rate, and sometimes the amount of data to process is less than the maximum amount. This corresponds to a fixed-throughput system whose workload requirements change on a block-to-block basis in a time varying fashion. Examples of this type of application include MPEG video processing and FIR filtering with a variable number of taps [68]. In the video processing example, the maximum workload corresponds to a scene change in the video sequence. In this case, the entire new frame of data requires processing since it is completely different from the previous frames. In the absence of scene changes, new frames of data may not differ significantly from the previous frame, so only a small section of the new frame requires processing. This case represents a reduced workload for the system. The workload of the system measures the amount of processing required for a given block of data, and the rate is simply the normalized processing frequency [68]. In a system without buffering, the lowest allowable rate equals the workload. If buffering is possible, then there are different strategies involving operation at different rates that can correctly perform the required processing on a block with a given workload by ensuring that the average rate equals the workload. There are many applications where workload varies with time [197], and policies for setting the rate based on incoming data have been explored [198][199][200].

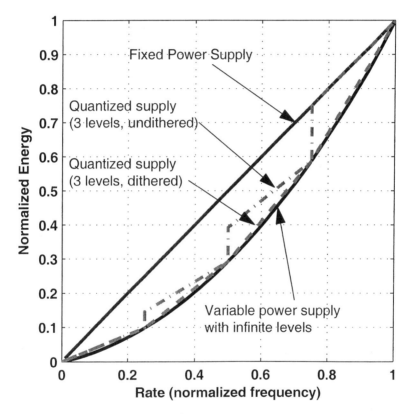

Fig. 9.7. Theoretical energy consumption versus rate for different power supply strategies [68]. (© 2006 IEEE)

Figure 9.7 shows four approaches to power supply management for reducing energy consumption when the workload varies [68]. It plots the required rate of the system versus the normalized energy required to process one generic block of data. The most straightforward method for saving energy when the workload decreases is to operate at the maximum rate until all of the required processing is complete and then to shutdown. This approach only requires a single power supply voltage (corresponding to full rate operation), and it results in linear energy savings. The fixed power supply curve in Figure 9.7 assumes ideal shutdown (i.e. - no shutdown power). A variable supply voltage with infinite allowable levels provides the optimum curve for reducing energy. This curve in Figure 9.7 corresponds to theoretically ideal DVS according to the model in [68] where velocity saturation is omitted. When velocity saturation occurs, the energy savings for ideal DVS increase because the performance does not decrease as quickly for the same change in V_{DD} [195].

Voltage Dithering

One method that avoids the problem of creating an infinite number of supply voltages is to use quantized supply voltages. In Figure 9.7, three levels of supply voltage quantization are used with two different policies. The undithered policy simply selects the lowest supply voltage for which the rate exceeds the desired rate, operates at that rate and voltage until all of the data in the block is processed, and then shuts down. This results in the stair-step energy characteristic. A better method is called voltage dithering [68]. The basic idea behind voltage dithering is to divide the computation of one block of data between operation at the quantized supply voltage and rate pairs that occur above and below the desired average rate. The energy profile for dithering between quantized voltage supplies linearly connects the quantized rate and energy pairs on the plot. Assuming that the desired rate of operation for a block, R_{BLOCK}, lies between two quantized rates, R_{LOW} and R_{HIGH}, then:

$$E_{BLOCK} = E_{LOW} \left(\frac{R_{HIGH} - R_{BLOCK}}{R_{HIGH} - R_{LOW}} \right) + E_{HIGH} \left(\frac{R_{BLOCK} - R_{LOW}}{R_{HIGH} - R_{LOW}} \right) \tag{9.1}$$

where E_{HIGH} and E_{LOW} are the normalized energies consumed for processing a block at R_{HIGH} and R_{LOW}, respectively.

Figure 9.8 shows an example comparing voltage dithering with fixed and variable supply approaches. This example uses two quantized voltages that provide full rate and half rate operation. For a desired rate of 0.6, the fixed supply approach operates at the highest rate and full power for 0.6 of the full block time (Figure 9.8(b)). Ideal variable voltage operation provides exactly 0.6 rate at the best possible energy (Figure 9.8(d)). Voltage dithering gives an average rate of 0.6 by operating for 20% of the block time at full rate and for 80% of the time at 0.5 rate. The resulting energy consumption is thus averaged between the two quantized points and falls on the connecting line (Figure 9.8(c)). This approach allows a good approximation of the optimum energy profile with less overhead.

Implementations of systems that use voltage dithering apply it monolithically to an entire chip. The system in [68] uses an on-chip variable DC-DC converter to dither the voltage supplied to the entire chip. A chip containing

9.2 Ultra-Dynamic Voltage Scaling 177

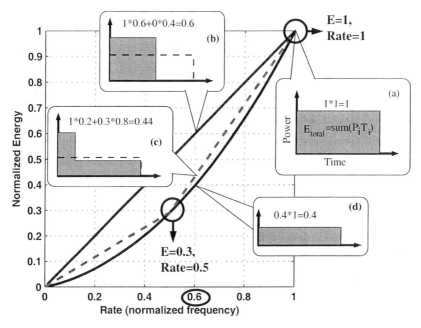

Fig. 9.8. Voltage dithering example for 0.6 rate normalized to full rate (a). Example shows fixed supply (b), voltage dithering (c), and ideal variable supply (d).

header switches was used in [195] to select the voltage supplied to a different chip (off-the-shelf processor). A similar off-chip voltage hopping approach is used in [201] for a zero V_T processor in fully depleted Silicon on Insulator (SOI). These implementations have shown the effectiveness of voltage dithering to save energy for high performance applications with variable workload.

Local Voltage Dithering

Applying voltage dithering at the local level provides several key advantages over previous chip-wide applications. We have proposed local voltage dithering (LVD) to improve upon chip-wide voltage dithering. This section discusses the advantages of LVD and describes a test-chip that demonstrates these improvements.

Previous chip-wide implementations using voltage dithering report that the transition between two different supply voltages takes hundreds of microseconds [68][195]. This prevents the system from achieving any energy savings for faster changes in the workload. Dividing up the power supply grid into local regions reduces the capacitance that must be switched when the voltage supplied to a local block needs to change. This allows for faster changes in

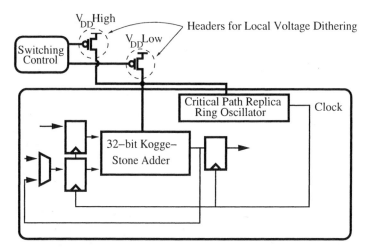

Fig. 9.9. Block diagram of voltage dithered adder and critical path replica using two local header switches for local voltage dithering (LVD). (© 2006 IEEE)

supply voltage with lower transitional energy and permits energy savings for changes in workload on the same timescale.

Chip-wide voltage dithering also restricts the extent to which varying workload may be leveraged because it must account for the highest workload from all of the blocks across the entire chip. For example, suppose a simple chip contains two large blocks. If one block has a workload of 0.9 and the other block has a workload of 0.2, then chip-wide voltage dithering must ensure that the block with the higher workload completes its work. Since both blocks share the dithered voltage supply, they both are forced to operate at the average rate of 0.9. Even if the less active block shuts down (e.g. clock gates) after completing its processing, it still uses more energy than if it could voltage dither based on its own workload. The energy savings that are lost by using chip-wide voltage dithering only increase with more blocks and wider differences between the maximum and minimum workloads. In contrast, LVD lets each block operate according to its own workload.

Our implementation of LVD uses embedded power switches (pMOS header devices) to toggle among a small number of voltage levels at the local block level. One advantage of this implementation approach is that the local dithering switches can be turned off to provide fine-grained power gating essentially for free.

9.2.2 UDVS Test Chip

We have implemented a test chip in 90nm bulk CMOS to demonstrate LVD and UDVS. This section describes the test chip architecture and provides measured results.

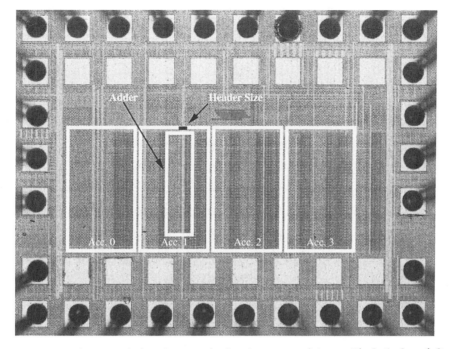

Fig. 9.10. Annotated die photograph showing accumulators with 0, 1, 2 and 3 headers. The size of one header is highlighted for reference. (© 2006 IEEE)

UDVS Test Chip Architecture

Figure 9.9 shows the primary block used for testing LVD on the test chip. The circuit of interest is a 32-bit Kogge-Stone adder that can be configured as an accumulator for testing. In this figure, two pMOS header switches select between a high supply voltage (V_{DDH}) and a low supply voltage (V_{DDL}) for the adder block. Other adders on the chip have different numbers of header devices. A critical path replica ring oscillator shares the same dithered voltage supply as the adder and sets the frequency of the clock based on the selected voltage. The die photo in Figure 9.10 shows the accumulators with different numbers of header switches used for testing, and the approximate area of a single header switch is highlighted for reference.

Placing a pMOS header switch in series with the power supply increases the delay of the circuit because of the voltage drop across the on resistance of the header. This effect is well-known and thoroughly analyzed in the context of power gating approaches such as multi-threshold CMOS (MTCMOS). Numerous methods for sizing such header devices are available, and most of them are designed to ensure that the circuit never exceeds some delay penalty relative to the circuit without any headers. The header switches on the test chip are sized to keep the delay penalty less than 10%.

9 System Examples

Figure 9.11 shows the architecture, and Figure 9.12 shows the circuit schematics for the adder block on the test chip. The inverse of the propagate and generate signals are calculated in the first stage, and these results are applied to the adder tree. Each reconverging point in the tree has a "dot operator" circuit that calculates the propagate and generate values for that stage. Each stage in our implementation is inverting, so the two flavors of dot operator are shown in the critical path schematic.

Measurements

Since UDVS scales the supply voltage from the full V_{DD} down to the optimum V_{DD} for minimum energy, a UDVS system must consist of circuits that can function in the sub-threshold region. This test chip uses static CMOS circuits to ensure robust sub-threshold operation. The adder blocks on the test chip operate to below 200mV. Figure 9.13 shows an oscilloscope plot of the adder on the 90nm test chip operating in sub-threshold at 300mV, just below the minimum energy voltage.

The minimum energy per operation point measured for the adder appears in Figure 9.14 at $V_{DD} = 330$mV ($f = 50$kHz) and 0.1pJ per addition for $25^{\circ}C$. Figure 9.14 also shows the measured effect of temperature on the total energy per cycle and leakage energy per cycle. An increase in temperature lowers the mobility of MOSFETs and decreases the threshold voltage according to:

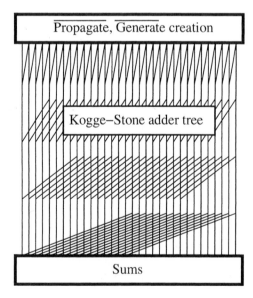

Fig. 9.11. Schematic of adder circuits. Kogge-Stone-based tree with inverting stages of dot operators (at each reconvergence of the tree). (© 2006 IEEE)

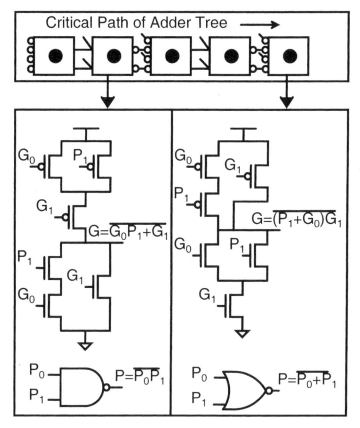

Fig. 9.12. Circuits for Kogge-Stone adder. Inverting stages of dot operators are in series along the critical path. Circuits do not require large stacks of transistors, which degrade sub-threshold operation. (© 2006 IEEE)

$\mu(T) = \mu(T_0)(\frac{T}{T_0})^{-M}$ and $V_T(T) = V_T(T_0) - KT$ [104]. For above-threshold operation, the decreased mobility dominates, and circuits slow down as they heat up. The leakage energy increases quickly with temperature for $V_{DD} > V_T$ because of the exponential dependence on the lower threshold voltage. In the sub-threshold region, however, the increased current also decreases the cycle delay, which causes the higher leakage currents to integrate over a shorter cycle time. As a result, the leakage energy does not change enough with temperature to greatly impact the optimum supply voltage. Figure 9.14 shows that the measured effect of temperature on the minimum energy point is small, validating the model in [99] and the analysis in Section 4.3.2. Figure 9.15 shows the measured frequency of one of the critical path ring oscillators on the test chip versus V_{DD} and temperature, confirming the increase of performance at higher temperatures in the sub-threshold region.

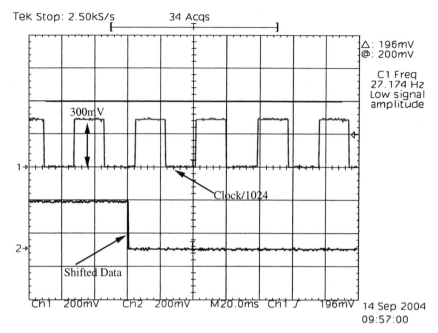

Fig. 9.13. Oscilloscope plot showing the clock and data from the 90nm test chip operating at 300mV, just below the minimum energy point. The adders functioned to 200mV. (© 2006 IEEE)

Figure 9.16 illustrates the savings that LVD provides for the adder block on the test chip when the rate varies. The dotted line shows operation at the highest rate followed by ideal shutdown. The solid line shows the measured energy versus rate for DVS assuming continuous voltage and frequency scaling. Selecting two rates from the curve, 1 and 0.5 in the figure, and operating for the correct fraction of time at each rate results in the dashed line that connects the quantized points, as described previously. A local block with three headers can achieve closer to optimum savings by selecting three rates and then dithering to connect those points on the plot.

While previously reported chip-wide approaches to voltage dithering have largely ignored the overhead energy of their schemes, we have investigated and measured the time and energy overhead of the LVD switching approach. Figure 9.17 shows the test circuit used to measure the delay overhead of LVD. While the adder runs a long accumulation, V_{DD} dithers to and from the higher rate. The oscilloscope plot in Figure 9.18 shows the divided ring oscillator output and the signal that selects the supply voltage (dither) for a dithering cycle between full and 0.5 rate. When the headers toggle V_{DD}, a counter gates the clock for a specified number of cycles to ensure settling at the new voltage. Checking the accumulated value verifies correct operation

9.2 Ultra-Dynamic Voltage Scaling   183

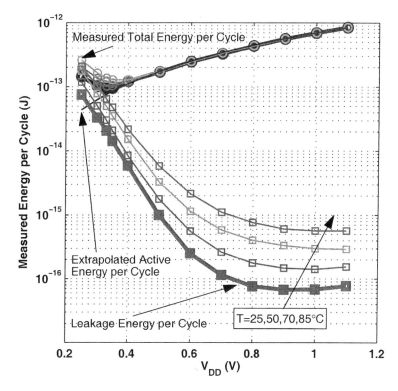

Fig. 9.14. Measured energy per cycle in the sub-threshold region for input activity of one. Minimum energy point occurs at 330mV (50kHz) and 0.1pJ per operation at 25°C. The optimum supply voltage is relatively insensitive to temperature variation. (© 2006 IEEE)

for every cycle. Measurements showed that the correct value was accumulated even with only 1/2 cycle (minimum possible using the test circuit) of clock gating for V_{DDL} above 0.6V, which corresponds to a rate of 0.04. Thus, even conservative settling times for this LVD implementation are on the order of a few cycles. This measurement confirms that LVD can adjust to fast changes in the workload of the local blocks.

In addition to timing overhead, there is energy overhead associated with the LVD approach. The buffer network and control circuits that drive the header switches consume energy every time they toggle the header switches to select a new supply voltage for the adder circuit. We can relate this overhead switching energy to the active switching energy of the adder block to determine its impact on overall energy savings from the LVD approach. To this end, we normalize the effective overhead switched capacitance of the control and buffer circuits, $C_{OVERHEAD}$, to the effective switched capacitance of the adder. The

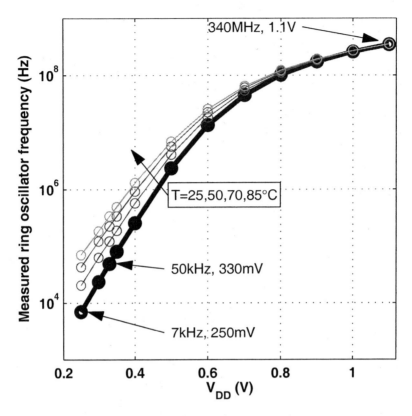

Fig. 9.15. Measured ring oscillator frequency versus V_{DD} and temperature. (© 2006 IEEE)

expression in Equation (9.2) shows the relation that must hold true in order for LVD to provide energy savings for a given transition.

$$NV_{DDH}^2 \geq NV_{DDL}^2 + C_{OVERHEAD}V_{DDH}^2 \qquad (9.2)$$

Solving (9.2) for N gives the number of cycles that must occur at V_{DDL} in order to make switching to V_{DDL} worthwhile for saving energy, as shown in (9.3).

$$N \geq \frac{C_{OVERHEAD}V_{DDH}^2}{V_{DDH}^2 - V_{DDL}^2} \qquad (9.3)$$

Measurements of the test chip show that $C_{OVERHEAD} = 3.7$ for the adder, so N is only 12 for the adder block with V_{DDH}=1.1V and V_{DDL}=0.9 (rate=0.5). Since the control circuits on the test chip are relatively simple, the overhead energy for more complicated control schemes, such as those that calculate the effective workload, has the effect of increasing N.

9.2 Ultra-Dynamic Voltage Scaling 185

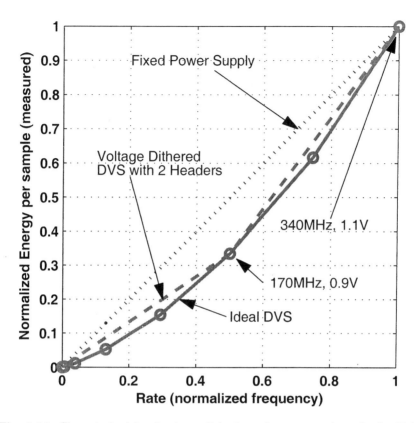

Fig. 9.16. Characterized local voltage dithering using measured results for 32-bit Kogge-Stone adder. (© 2006 IEEE)

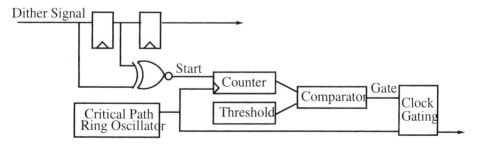

Fig. 9.17. Circuit for measuring timing overhead of LVD that gates the clock at a V_{DD} transition for a given number of cycles. The duration of this clock gating is decreased until the circuit fails. (© 2006 IEEE)

186   9 System Examples

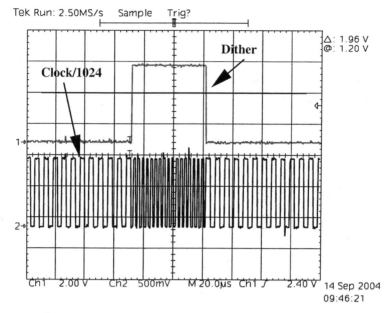

Fig. 9.18. Oscilloscope plot showing the system clock while dithering between rate 0.5 (170MHz) and rate 1 (340MHz). Measurements show correct accumulation at both transitions even no clock gating (see Figure 9.17). (© 2006 IEEE)

9.2.3 UDVS System Considerations

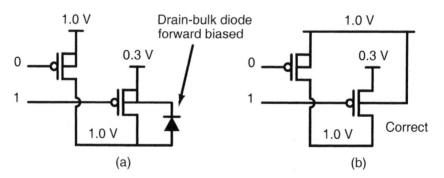

Fig. 9.19. For UDVS, the bulk connections of the pMOS header switches have to connect to the highest supply voltage. (© 2006 IEEE)

The discussion to this point has assumed that the varying rate remains above roughly a few percent. As previously mentioned, sub-threshold operation has proven to minimize energy for low performance applications. While scaling to sub-threshold is rarely advantageous for full processors [101], local blocks or special applications that require brief periods of high performance spend significant amounts of time operating at effective rates that are orders of magnitude below one. Examples of these applications include micro-sensor nodes, medical devices, wake-up circuitry for processors, and local blocks on active processors. When performance is non-critical, energy is minimized by operating at the minimum energy point that occurs because of increased leakage energy at low frequency and then shutting down if there is more timing slack.

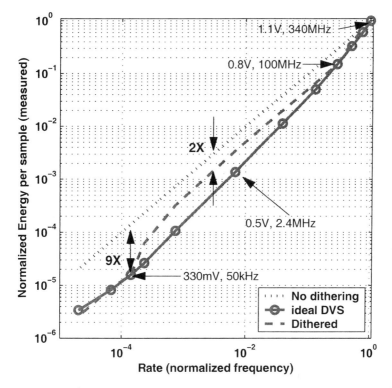

Fig. 9.20. UDVS) using two headers with one variable DC-DC converter or using three headers (c.f. Figure 9.23). (© 2006 IEEE)

Since LVD works well for high speed operation and operating at the minimum energy point is optimal for low performance situations, we propose ultra-dynamic voltage scaling (UDVS) using local power switches [196]. This approach uses local headers to perform LVD when high performance is necessary and selects a low voltage for sub-threshold operation at the minimum energy point whenever performance is not critical. As with LVD, all of the headers for a given block can turn off when the block is idle to conserve standby power using power gating. Since the power switches in the UDVS approach connect to different voltages that can differ substantially, they must be configured carefully to prevent forward biasing the junction diodes. Figure 9.19 shows that connecting the bulk terminal of the header transistors to the source can forward bias the drain-bulk diode when the V_{DDL} switch is off. One solution to this problem is to tie the bulk of all of the header switches to the highest supply voltage as in Figure 9.19(b).

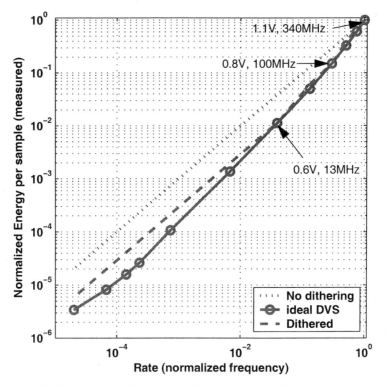

Fig. 9.21. Different choice of dithered voltages for closer fit over the higher range of V_{DD}. (© 2006 IEEE)

Figure 9.20 provides one example of measured UDVS characteristics for the adder. In this example, dithered voltages are chosen at 1.1V, 0.8V, and 0.33V, which is the optimum voltage for minimum energy. When the adder block is performing operations with no timing deadline, it functions at the

9.2 Ultra-Dynamic Voltage Scaling

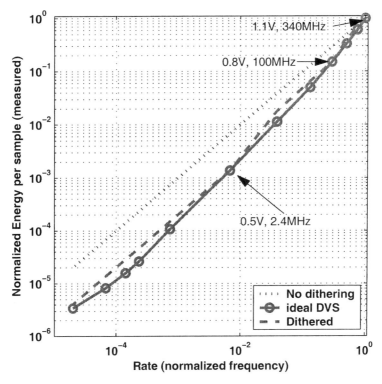

Fig. 9.22. Different choice of dithered voltages for closer fit over the entire range of V_{DD}. (© 2006 IEEE)

minimum energy point at 50kHz and saves 9X the energy versus the ideal shutdown scenario. When performance becomes important, the adder dithers between 1.1V and 0.8V within 30% of the optimal energy consumption while adjusting for variations in the rate above 0.1. It was shown in [202] that significant extra savings are available if the selected dithered rates match to the prominent average rates in the data. This brings the dithered curve closer to the optimum DVS curve for the common cases. Figure 9.21 and Figure 9.22 show two additional examples in which the supply voltages are chosen for different scenarios. For a system whose rate requirements vary evenly over the full range, the voltage choices in Figure 9.21 provide a better match to the ideal energy profile, but the minimum energy per operation is not achievable. If performance constraints prevent a system from ever operating at the minimum energy point, the supply voltage can be adjusted to higher voltages to achieve near optimal energy operation over the range of higher rates (Figure 9.22).

Figure 9.23 shows two options for implementing the power supplies and headers in a UDVS system. The straightforward option is to distribute three

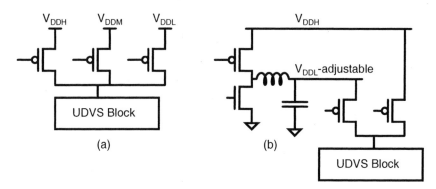

Fig. 9.23. Options for UDVS headers at the system level. (© 2006 IEEE)

supply voltages around the chip and to use three header switches at each block. The voltages V_{DDH}, V_{DDM}, and V_{DDL} can be selected based on the system workload statistics as we described above. The only advantage to using more than three power supplies is to pin the UDVS energy profile to the ideal variable supply profile in more places. The right-hand diagram in Figure 9.23 offers a second option for implementing UDVS. When transitions between high and low performance mode are infrequent and the amount of time in between transitions is long, two header switches may be paired with one adjustable DC-DC converter for the same functionality. For example, during high speed operation, the headers dither between 1.1V and 0.8V. When the rare transition to low speed occurs, the DC-DC converter switches V_{DDL} to 0.35V so that all of the blocks can operate near their minimum energy points.

For applications where some blocks operate in sub-threshold while others are at higher voltages, special interfacing circuits are required at the low voltage region to high voltage region interface. The type of level converters to be used will depend on how the block interfaces to surrounding blocks. Ample previous work on level converter circuits offers many choices for implementing the required interfaces. In a full UDVS system with multiple blocks, each block has its own header devices so that it can voltage dither based on its individual workload. Communication among blocks occurs along a bus, which might be asynchronous to account for different operating frequencies, and level converters interface to the bus as needed.

A
Acronyms

6T	six transistor
AOI	And/Or/Invert
BL	bitline
BW	Baugh-Wooley
CDF	Cumulative Distribution Function
CE	Constant Electric Field
CMOS	Complementary MOSFET
CV	Constant Voltage
DIBL	Drain-Induced Barrier Lowering
DIO	Data Input/Output
DLMS	Delayed Least Mean Square
DSM	Deep Sub-Micron
DSP	Digital Signal Processor
DVS	Dynamic Voltage Scaling
EDP	energy-delay product
EKV	Enz, Krummenacher, and Vittoz
FBB	Forward Body Bias
FFT	Fast Fourier Transform
FIR	Finite Impulse Response
FO4	Fan-Out of 4
GIDL	Gate-Induced Drain Leakage
IC	Integrated Circuit
ITRS	International Technology Roadmap for Semiconductors
LFSR	Linear Feedback Shift Register
LVD	Local Voltage Dithering
MCU	Microcontroller Units
MTCMOS	multi-threshold CMOS
nMOS	n-type MOSFET
NTRS	National Technology Roadmap for Semiconductors
PDP	power delay product
P-nMOS	Pseudo-NMOS

pMOS	p-type MOSFET
QCV	Quasi-Constant Voltage
RO	ring oscillator
RBB	Reverse Body Bias
RBL	read bitline
RFID	Radio Frequency Identification
RVFFT	Real-Valued FFT
RWL	read wordline
SER	Soft Error Rate
SIA	Semiconductor Industry Association
SNM	Static Noise Margin
SOI	Silicon on Insulator
SRAM	Static Random Access Memory
SS	Strong nMOS, Strong pMOS
SW	Strong nMOS, Weak pMOS
TG	Transmission Gate
TI	Texas Instruments
TT	Typical nMOS, Typical pMOS
UDVS	Ultra-Dynamic Voltage Scaling
ULP	Ultra Low Power
VCO	voltage-controlled oscillator
VLIW	Very-Long Instruction Word
VTC	Voltage Transfer Characteristic
WL	wordline
WS	Weak nMOS, Strong pMOS
WW	Weak nMOS, Weak pMOS

References

1. R. M. Swanson and J. D. Meindl, "Ion-Implanted Complementary MOS Transistors in Low-Voltage Circuits," *IEEE Journal of Solid-State Circuits*, vol. 7, no. 2, pp. 146–153, April. 1972.
2. A. Wang, "An ultra-low voltage FFT processor using energy-aware techniques," Ph.D. dissertation, Massachusetts Institute of Technology, 2003.
3. B. Calhoun, "Low Energy Digital Circuit Design Using Sub-threshold Operation," Ph.D. dissertation, Massachusetts Institute of Technology, 2005.
4. A. Cerpa, J. Elson, D. Estrin, L. Girod, M. Hamilton, and J. Zhao, "Habitat Monitoring: Application Driver for Wireless Communications Technology," in *Proceedings of the ACM SIGCOMM Workshop on Data Communications in Latin America and the Caribbean*, 2001, pp. 20–41.
5. A. Mainwaring, J. Polastre, R. Szewczyk, D. Culler, and J. Anderson, "Wireless Sensor Networks for Habitat Monitoring," in *ACM International Workshop on Wireless Sensor Networks and Applications (WSNA)*, 2002, pp. 88–97.
6. E. Biagioni and K. Bridges, "The Application of Remote Sensor Technoloy to Assist the Recovery of Rare and Endangered Species," *Special Issue on Distributed Sensor Networks for the International Journal of High Performance Computing Applications*, vol. 16, no. 3, pp. 315–324, Aug. 2002.
7. L. Schwiebert, S. Gupta, and J. Weinmann, "Research Challenges in Wireless Networks of Biomedical Sensors," in *Mobile Computing and Networking*, 2001, pp. 151–165.
8. K. Chintalapudi, E. Johnson, and R. Govindan, "Structural Damage Detection Using Wireless Sensor-Actuator Networks," in *Proceedings of the IEEE International Symposium on Intelligent Control*, 2005, pp. 322–327.
9. R. Measures, "Fiber Optic Structural Monitoring of Bridges," in *Proceedings of the IEEE Instrumentation and Measurement Technology Conference*, 1997, pp. 600–602.
10. A. Basharat, N. Catbas, and M. Shah, "A Framework for Intelligent Sensor Network with Video Camera for Structural Health Monitoring of Bridges," in *Proceedings of the IEEE Conference on Pervasive Computing and Communications (PerCom)*, 2005, pp. 385–389.
11. R. Weinstein, "RFID: a technical overview and its application to the enterprise," *IT Professional*, vol. 7, no. 3, pp. 27–33, May-June 2005.

12. S. Roundy, P. Wright, and J. Rabaey, "A Study of Low Level Vibrations as a Power Source for Wireless Sensor Nodes," *Computer Communications*, vol. 26, no. 11, pp. 1131–1144, 2003.
13. R. Hahn and H. Reichl, "Batteries and power supplies for wearable and ubiquitous computing," in *Proc. 3es Int. Symp. Wearable Computers*, 1999, pp. 168–169.
14. B. H. Calhoun, D. Daly, N. Verma, D. Finchelstein, D. D. Wentzloff, A. Wang, S.-H. Cho, and A. Chandrakasan, "Design Considerations for Ultra-Low Energy Wireless Microsensor Nodes," *IEEE Trans. on Computers*, vol. 54, no. 6, pp. 727–740, 2005.
15. H. Kulah and K. Najafi, "An Electromagnetic Micro Power Generator for Low-Frequency Environmental Vibrations," in *Proceedings of the IEEE International Conference for Micro Electro Mechanical Systems (MEMS)*, Jan. 2004, pp. 237–240.
16. S. Meninger, J. Mur-Miranda, R. Amirtharajah, A. Chandrakasan, and J. Lang, "Vibration-to-electric Energy Conversion," *IEEE Transactions on Very Large Scale Integration (VLSI) Systems*, vol. 9, no. 1, pp. 64–76, Feb. 2001.
17. H. Bottner, J. Nurnus, A. Gavrikov, G. Kuhner, M. Jagle, C. Kunzel, D. Eberhard, G. Plescher, A. Schubert, and K.-H. Schlereth, "New Thermoelectric Components Using Microsystems Technologies," *IEEE/ASME Journal of Microelectromechanical Systems*, vol. 13, no. 3, pp. 414–420, 2004.
18. *Panasonic Solar Cells Technical Handbook '98/'99*, Matsushita Battery Industrial Co., Ltd., Aug. 1998.
19. J. Kahn, R. Katz, and K. Pister, "Next Century Challenges: Mobile Networking for Smart Dust," in *Proceedings of the ACM MobiCom*, 1999, pp. 271–278.
20. M. Hempstead, N. Tripathi, P. Mauro, G. Wei, and D. Brooks, "An ultra low power system architecture for sensor network applications," in *International Symposium on Computer Architecture*, 2005, pp. 208–219.
21. C. G. B. Garrett and W. H. Brattain, "Physical Theory of Semiconductor Surfaces," *Physical Review*, vol. 99, p. 376, 1955.
22. S. M. Sze, *Physics of Semiconductor Devices*, 1^{rst} ed. John Wiley & Sons, 1969.
23. http://invention.smithsonian.org/centerpieces/quartz/inventors/swissinvent.html.
24. F. M. Wanlass and C. T. Sah, "Nanowatt Logic Using Field-Effect Metal-Oxide-Semiconductor Triodes," in *IEEE International Solid-State Circuits Conference (ISSCC) Digest of Technical Papers*, vol. VI, Philadelphia, Febr. 1963, pp. 32–33.
25. M. B. Barron, "Low Level Currents in Insulated Gate Field Effect Transistors," *Solid-State Electronics*, vol. 15, pp. 293–302, 1972.
26. R. R. Troutman and S. N. Chakravarti, "Subthreshold Characteristics of Insulated-Gate Field-Effect Transistors ," *IEEE Trans. Circuit Theory*, vol. 20, no. 6, pp. 659–665, Nov. 1973.
27. R. R. Troutman, "Subthreshold Slope for Insulated Gate Field-Effect Transistors," *IEEE Trans. Electron Devices*, vol. 22, pp. 1049–1051, Nov. 1975.
28. T. Masuhara, J. Etoh, and M. Nagata, "A Precise MOSFET Model for Low-Voltage Circuits," *IEEE Trans. Electron Devices*, vol. 21, no. 6, pp. 363–371, June 1974.

29. R. W. J. Barker, "Small-Signal Subthreshold Model for IGFET's," *El. Letters*, vol. 12, pp. 260–262, May 1976.
30. E. Vittoz and J. Fellrath, "New Analog CMOS IC's Based on Weak Inversion Operation," in *Proceedings of the European Solid-State Circuits Conference*, Toulouse, Sept. 1976, pp. 12–13.
31. ——, "CMOS Analog Integrated Circuits Based on Weak Inversion Operation," *IEEE Journal of Solid-State Circuits*, vol. 12, no. 3, pp. 224–231, June 1977.
32. E. A. Vittoz and O. Neyroud, "A Low-Voltage CMOS Bandgap Reference," *IEEE Journal of Solid-State Circuits*, vol. 14, no. 3, pp. 573–577, June 1979.
33. J. J. Ebers and J. L. Moll, "Large-Signal Behavior of Junction Transistors," *Proc. IRE*, vol. 42, no. 12, pp. 1761–1772, Dec. 1954.
34. J. Fellrath and E. Vittoz, "Small Signal Model of MOS Transistors in Weak Inversion," in *Proc. Journées d'Electronique '77*, EPF-Lausanne, 1977, pp. 315–324.
35. Y. P. Tsividis and R. W. Ulmer, "A CMOS Voltage Reference," *IEEE Journal of Solid-State Circuits*, vol. 13, no. 6, pp. 774–778, Dec. 1978.
36. O. H. Schade, "BiMOS Micropower IC's," *IEEE Journal of Solid-State Circuits*, vol. 13, no. 6, pp. 791–798, Dec. 1978.
37. E. Vittoz, "Quartz Oscillators for Watches," in *Proc. 10th International Congress of Chronometry*, Geneva, 1979, pp. 131–140.
38. ——, "Micropower Switched-Capacitor Oscillator," *IEEE Journal of Solid-State Circuits*, vol. 14, pp. 622–624, June 1979.
39. E. Vittoz and F. Krummenacher, "Micropower SC Filters in Si-Gate CMOS technology," in *Proc. ECCTD'80*, Warshaw, 1980, pp. 61–72.
40. F. Krummenacher, E. Vittoz, and M. Degrauwe, "Class AB CMOS Amplifier for Micropower SC Filters," *El. Letters*, vol. 17, no. 13, pp. 433–435, June 1981.
41. E. Vittoz, "Microwatt SC Circuit Design," *Electrocomponent Science and Technology*, vol. 9, pp. 263–273, 1982.
42. M. Degrauwe, J. Rijmenants, E. Vittoz, and H. De Man, "Adaptive Biasing CMOS Amplifiers," *IEEE Journal of Solid-State Circuits*, vol. 17, pp. 522–528, June 1982.
43. F. Krummenacher, "Micropower Switched Capacitor Biquadratic Cell," *IEEE Journal of Solid-State Circuits*, vol. 17, pp. 507–512, June 1982.
44. M. Degrauwe, E. Vittoz, and I. Verbauwhede, "A Micropower CMOS Instrumentation Amplifier," *IEEE Journal of Solid-State Circuits*, vol. 20, pp. 805–807, June 1985.
45. E. Vittoz, "Micropower Techniques," in *Design of MOS VLSI Circuits for Telecommunications*, Y. Tsividis and P. Antognetti, Eds. Prentice-Hall, 1985.
46. E. Vittoz, M. Degrauwe, and S. Bitz, "High-performance Crystal Oscillator Circuits: Theory and Applications," *IEEE Journal of Solid-State Circuits*, vol. 23, no. 3, pp. 774–783, June 1988.
47. H. Oguey and S. Cserveny, "Modèle du transistor MOS valable dans un grand domaine de courants," *Bulletin ASE*, vol. 73, no. 3, pp. 113–116, 1982, in French.
48. ——, "MOS Modelling at Low Current Density," ESAT," Summer Course on Process and Device Modelling, June 1983.
49. C. Enz, F. Krummenacher, and E. Vittoz, "An Analytical MOS Transistor Model Valid in All Regions of Operation and Dedicated to Low-Voltage and Low-Current Applications," *Special issue of the Analog Integrated Circuits and*

Signal Processing Journal on Low-Voltage and Low-Power Design, vol. 8, pp. 83–114, July 1995.
50. C. Mead, *Analog VLSI and Neural Systems*. Addison-Wesley, 1989.
51. M. A. Maher and C. Mead, "A Physical Charge-Controlled Model for the MOS Transistors," in *Advanced Research in VLSI, Proc. of the 1987 Stanford Conference*, P. Losleben, Ed. Cambridge, MA: MIT Press, 1987.
52. C. Enz and E. Vittoz, *Charge-Based MOS Transistor Modeling: The EKV Model for Low-Power and RF IC Design*. John Wiley & Sons, 2006.
53. E. Vittoz, "Weak Inversion for Ultimate Low-Power Logic," in *Low-Power Electronics Design*, C. Piguet, Ed. CRC Press, 2005.
54. S. Sze, "Semiconductor device development in the 1970's and 1980's - A perspective," *Proceedings of the IEEE*, vol. 69, no. 9, pp. 1121 – 1131, Sept 1981.
55. "International technology roadmap for semiconductors," 2003. [Online]. Available: http://public.itrs.net
56. C. Svensson, "Forty Years of Feature-Size Predictions (1962-2002)," in *IEEE International Solid-State Circuits Conference (ISSCC) Digest of Technical Papers*, 2003.
57. J. Swanson, "Physical versus Logical Coupling in Memory Systems," *IBM Journal*, vol. 4, pp. 305–310, 1960.
58. R. Dennard, F. Gaensslen, V. Rideout, E. Bassous, and A. LeBlanc, "Design of ion-implanted MOSFET's with very small physical dimensions," *IEEE Journal of Solid-State Circuits*, vol. 9, no. 5, pp. 256– 268, Oct 1974.
59. J. Meindl, "Theoretical, practical and analogical limits in ULSI," in *International Electron Devices Meeting (IEDM) Digest of Technical Papers*, vol. 29, 1983, pp. 8 – 13.
60. P. Chatterjee, W. Hunter, T. Holloway, and Y. Lin, "Impact of scaling laws on the choice of n-channel or p-channel for MOS VLSI," *IEEE Electron Device Letters*, vol. 1, no. 10, pp. 220 – 223, October 1980.
61. M. Kakumu, M. Kinugawa, K. Hashimoto, and J. Matsunaga, "Power supply voltage for future CMOS VLSI in half and sub micrometer," in *International Electron Devices Meeting (IEDM) Digest of Technical Papers*, vol. 32, 1986, pp. 399– 402.
62. R. Allmon, B. Benschneider, M. Callander, L. Chao, D. Dever, J. Farrell, N. Fitzgerald, J. Grodstein, S. Hassoun, L. Hudepohl, D. Kravitz, J. Lundberg, R. Marcello, S. Marino, J. Pickholtz, R. Preston, M. Richesson, S. Samudrala, and D. Sanders, "System, process, and design implications of a reduced supply voltage microprocessor," in *IEEE International Solid-State Circuits Conference (ISSCC) Digest of Technical Papers*, vol. 33, 1990, pp. 48–49.
63. A. Chandrakasan, A. Burstein, and R. Brodersen, "A low power chipset for portable multimedia applications," in *IEEE International Solid-State Circuits Conference (ISSCC) Digest of Technical Papers*, vol. 37, Feb 1994, pp. 82–83.
64. J. Burr and J. Shott, "A 200mV self-testing encoder/decoder using Stanford ultra-low-power CMOS," in *IEEE International Solid-State Circuits Conference (ISSCC) Digest of Technical Papers*, vol. 37, Feb 1994, pp. 84–85.
65. T. Kuroda, T. Fujita, S. Mita, T. Nagamatu, S. Yoshioka, F. Sano, M. Norishima, M. Murota, M. Kako, M. Kinugawa, M. Kakumu, and T. Sakurai, "A 0.9V 150MHz 10mW 4mm2 2-D discrete cosine transform core processor with variable-threshold-voltage scheme," in *IEEE International Solid-State Circuits Conference (ISSCC) Digest of Technical Papers*, vol. 39, Feb 1996, pp. 166 – 167.

66. S. Mutoh, S. Shigematsu, Y. Matsuya, H. Fukuda, and J. Yamada, "A 1V multi-threshold voltage CMOS DSP with an efficient power management technique for mobile phone application," in *IEEE International Solid-State Circuits Conference (ISSCC) Digest of Technical Papers*, vol. 39, Feb 1996, pp. 168 – 169.
67. W. Lee, P. Landman, B. Barton, S. Abiko, H. Takahashi, H. Mizuno, S. Muramatsu, K. Tashiro, M. Fusumada, L. Pham, F. Boutaud, G. Gallo, H. Tran, C. Lemonds, A. Shih, M. Nandakumar, and B. Eklund, "A 1V DSP for wireless communications," in *IEEE International Solid-State Circuits Conference (ISSCC) Digest of Technical Papers*, vol. 40, Feb 1997, pp. 92 – 93.
68. V. Gutnik and A. Chandrakasan, "Embedded Power Supply for Low-Power DSP," *IEEE Transactions on Very Large Scale Integration (VLSI) Systems*, vol. 5, no. 4, pp. 425–435, Dec. 1997.
69. L. Clark, E. Hoffman, M. Schaecher, M. Biyani, D. Roberts, and Y. Liao, "A scalable performance 32b microprocessor," in *IEEE International Solid-State Circuits Conference (ISSCC) Digest of Technical Papers*, vol. 44, Feb 2001, pp. 230 – 231.
70. M. Horowitz, E. Alon, D. Patil, S. Naffziger, R. Kumar, and K. Bernstein, "Scaling, Power, and the Future of CMOS," in *International Electron Devices Meeting (IEDM) Digest of Technical Papers*, 2005.
71. R. Keyes and T. Watson, "On power dissipation in semiconductor computing elements," *Proc. IRE (Corresp.)*, vol. 50, p. 2485, Dec 1962.
72. J. Meindl and R. Swanson, "Potential improvements in power-speed performance of digital circuits," in *Proceedings of the IEEE*, vol. 59, no. 5, May 1971, pp. 815 – 816.
73. R. Swanson, "Complementary MOS Transistors in Micropower Circuits," Ph.D. dissertation, Stanford University, 1974.
74. R. Keyes, "Fundamental limits in Digital Information Processing," in *Proceedings of the IEEE*, vol. 69, no. 2, Feb 1981.
75. J. Burr and A. Peterson, "Energy Considerations in Multichip-Module Based Multiprocessors," in *IEEE International Conference on Computer Design (ICCD) Digest of Technical Papers*, 1991, pp. 593–600.
76. J. Burr, "Cryogenic Ultra Low Power CMOS," in *International Symposium on Low-Power Electronics and Design (ISLPED) Digest of Technical Papers*, 1995, pp. 82–83.
77. G. Schrom and S. Selberherr, "Ultra-Low-Power CMOS Technologies," in *International Semiconductor Conference (CAS) Digest of Technical Papers*, 1996, pp. 237–246.
78. A. Bryant, J. Brown, P. Cottrell, M. Ketchen, J. Ellis-Monaghan, and J. Nowak, "Low-Power CMOS at Vdd=4kT/q," in *Device Research Conference*, June 2001, pp. 22–23.
79. G. Ono and M. Miyazaki, "Threshold-voltage Balance for Minimum Supply Operation," in *Symposium On VLSI Circuits Digest of Technical Papers*, 2002.
80. J. Kao, M. Miyazaki, and A. Chandrakasan, "A 175-mV Multiply-Accumulate Unit Using an Adaptive Supply Voltage and Body Bias Architecture," *IEEE Journal of Solid-State Circuits*, vol. 37, no. 11, pp. 1545–1554, Nov. 2002.
81. J. Burr and A. Peterson, "Ultra Low Power CMOS Technology," in *3rd NASA Symposium on VLSI Design*, 1991, pp. 4.2.1–4.2.13.

82. R. Gonzalez, B. Gordon, and M. Horowitz, "Supply and Threshold Voltage Scaling for Low Power CMOS," *IEEE Journal of Solid-State Circuits*, vol. 32, no. 8, pp. 1210–1216, Aug. 1997.
83. M. Stan, "Optimal Voltages and Sizing for Low Power," in *International Conference on VLSI Design (VLSI-Design) Digest of Technical Papers*, 1999, pp. 428–433.
84. K. Nose and T. Sakurai, "Optimization of V_{DD} and V_{TH} for Low-Power and High-Speed Applications," in *ACM/IEEE Design Automation Conference (DAC) Digest of Technical Papers*, 2000, pp. 469–474.
85. A. Bhavnagarwala, B. Austin, K. Bowman, and J. Meindl, "A Minimum Total Power Methodology for Projecting Limits on CMOS GSI," *IEEE Transactions on Very Large Scale Integration (VLSI) Systems*, vol. 8, no. 3, pp. 235–251, 2000.
86. R. Brodersen, M. Horowitz, D. Markovic, B. Nikolic, and V. Stojanovic, "Methods for True Power Minimization," in *IEEE International Conference on Computer-Aided Design (ICCAD) Digest of Technical Papers*, 2002, pp. 35–42.
87. E. Vittoz, B. Gerber, and F. Leuenberger, "Silicon-Gate CMOS Frequency Divider for the Electronic Wrist Watch," *IEEE Journal of Solid-State Circuits*, vol. SC-7, pp. 100–104, 1972.
88. E. Vittoz, private communication.
89. R. Lyon and C. Mead, "An analog electronic cochlea," in *IEEE Transactions on Acoustics, Speech, and Signal Processing*, vol. 36, no. 7, July 1988, pp. 1119–1134.
90. C. H. Kim, H. Soeleman, and K. Roy, "Ultra-Low-Power DLMS Adaptive Filter for Hearing Aid Applications," *IEEE Transactions on Very Large Scale Integration (VLSI) Systems*, vol. 11, no. 4, pp. 716–730, Aug. 2003.
91. B. Paul, H. Soeleman, and K. Roy, "An 8 X 8 sub-threshold digital CMOS carry save array multiplier," in *Proceedings of the European Solid-State Circuits Conference*, September 2001.
92. M. Deen, H. Kazemeini, and S. Naseh, "Ultra-low Power VCOs - Performance Characteristics and Modeling," in *IEEE Internationl Caracas Conference on Devices, Circuits and Systems Digest of Technical Papers*, Apr. 2002, pp. C033-1 to C033-8.
93. B. H. Calhoun and A. Chandrakasan, "A 256kb Sub-threshold SRAM in 65nm CMOS," in *IEEE International Solid-State Circuits Conference (ISSCC) Digest of Technical Papers*, vol. 49, 2006, pp. 628–629.
94. A. Wang, A. Chandrakasan, and S. Kosonocky, "Optimal Supply and Threshold Scaling for Sub-threshold CMOS Circuits," in *IEEE Computer Society Annual Symposium on VLSI*, Apr. 2002, pp. 7–11.
95. V. De, Y. Ye, A. Keshavarzi, S. Narendra, J. Kao, D. Somasekhar, R. Nair, and S. Borkar, "Techniques for Leakage Power Reduction," in *Design of High-Performance Microprocessor Circuits*, A. Chandrakasan, W. Bowhill, and F. Fox, Eds. IEEE Press, 2001, ch. 3, pp. 46–62.
96. K. Roy, S. Mukhopadhyay, and H. Mahmoodi-Meimand, "Leakage Current Mechanisms and Leakage Reduction Techniques in Deep-Submicrometer CMOS Circuits," *Proceedings of the IEEE*, vol. 91, no. 2, pp. 305–327, Feb. 2003.

97. M. Miura-Mattausch, M. Suetake, J. Mattausch, S. Kumashiro, and N. Shigyo, "Physical Modeling of the Reverse-Short-Channel Effect for Circuit Simulation," *IEEE Transactions on Electron Devices*, vol. 48, no. 10, pp. 2449–2452, Oct. 2001.
98. A. Ono, R. Ueno, and I. Sakai, "TED Control Technology for Suppression of Reverse Narrow Channel Effect in $0.1\mu m$ MOS Devices," in *International Electron Devices Meeting (IEDM) Digest of Technical Papers*, 1997, pp. 227–230.
99. B. H. Calhoun and A. Chandrakasan, "Characterizing and Modeling Minimum Energy Operation for Subthreshold Circuits," in *International Symposium on Low-Power Electronics and Design (ISLPED) Digest of Technical Papers*, 2004, pp. 90–95.
100. B. Calhoun, A. Wang, and A. Chandrakasan, "Modeling and Sizing for Minimum Energy Operation in Subthreshold Circuits," *IEEE Journal of Solid-State Circuits*, vol. 40, no. 9, pp. 1778–1786, Sept. 2005.
101. B. Zhai, D. Blaauw, D. Sylvester, and K. Flautner, "Theoretical and Practical Limits of Dynamic Voltage Scaling," in *ACM/IEEE Design Automation Conference (DAC) Digest of Technical Papers*, 2004, pp. 868–873.
102. R. Corless, G. Gonnet, D. Hare, D. Jeffrey, and D. Knuth, "On the Lambert W Function," *Advances in Computational Mathematics*, vol. 5, pp. 329–359, 1996.
103. A. Wang and A. Chandrakasan, "A 180mV FFT Processor Using Sub-threshold Circuit Techniques," in *IEEE International Solid-State Circuits Conference (ISSCC) Digest of Technical Papers*, 2004, pp. 292–293.
104. A. Bellaouar, A. Fridi, M. Elmasry, and K. Itoh, "Supply Voltage Scaling for Temperature Insensitive CMOS Circuit Operation," *IEEE Transactions on Circuits and Systems*, vol. 45, no. 3, pp. 415–417, 1998.
105. A. Chandrakasan, S. Sheng, and R. Brodersen, "Low-Power CMOS Digital Design," *IEEE Journal of Solid-State Circuits*, vol. 27, no. 4, pp. 473–484, Apr. 1992.
106. C. Enz and E. Vittoz, "CMOS Low-Power Analog Circuit Design," in *Emerging Technologies*, R. Cavin and W. Liu, Eds. IEEE Operations Center, 1996.
107. S. M. Sze, *Semiconductor Devices: Physics and Technology*, 2nd ed. John Wiley & Sons, 1981.
108. Y. Tsividis, *Operation and Modeling of the MOS Transistor*, 2nd ed. New-York: Mc-Graw-Hill, 1999.
109. E. Vittoz, C. Enz, and F. Krummenacher, "A Basic Property of MOS Transistors and its Circuit Implications," in *Workshop on Compact Modeling at the International Conference on Modeling and Simulation of Microsystems*, vol. 2. San Francisco: Computational Publications, February 2003, pp. 246–249.
110. E. Vittoz, "Micropower Techniques," in *Design of Analog-Digital VLSI Circuits for Telecommunications and Signal Processing*, J. Franca and Y. Tsividis, Eds. Prentice-Hall, 1994.
111. E. Vittoz and X. Arreguit, "Linear Networks Based on Transistors," *El. Letters*, vol. 29, Feb. 1993.
112. E. Vittoz, "Pseudo-Resistive Networks and Their Applications to Analog Collective Computation," in *Proc. of MicroNeuro 97*, Dresden, 1997, pp. 163–173.
113. K. Bult and G. Geelen, "An Inherently Linear and Compact MOST-Only Current Division Technique," *IEEE Journal of Solid-State Circuits*, vol. 27, pp. 1730–1735, December 1992.

114. J. Fellrath, "Shot Noise Behaviour of Subthreshold MOS Transistors," *Revue de Physique Applique*, vol. 13, pp. 719–723, Dec. 1978.
115. E. A. Vittoz, "The Design of High-Performance Analog Circuits on Digital CMOS Chips," *IEEE Journal of Solid-State Circuits*, vol. 20, no. 3, pp. 657–665, June 1985.
116. M. J. M. Pelgrom, A. C. J. Duinmaijer, and A. P. G. Welbers, "Matching Properties of MOS Transistors," *IEEE Journal of Solid-State Circuits*, vol. 24, no. 5, pp. 1433–1440, Oct. 1989.
117. E. Vittoz, "Low-Power Design: Ways to Approach the Limits," in *IEEE International Solid-State Circuits Conference (ISSCC) Digest of Technical Papers*, 1994, pp. 14–18.
118. C. Piguet, "Design of Low-Power Libraries," in *International Conference on Electronics, Circuits and Systems (ICECS) Digest of Technical Papers*, vol. 2, 1998, pp. 175–180.
119. C. Piguet, J.-M. Masgonty, S. Cserveny, C. Arm, and P.-D. Pfister, "Low-Power Low-Voltage Library Cells and Memories," in *International Conference on Electronics, Circuits and Systems (ICECS) Digest of Technical Papers*, 2001, pp. 1521–1524.
120. B. H. Calhoun, A. Wang, and A. Chandrakasan, "Device Sizing for Minimum Energy Operation in Subthreshold Circuits," in *Custom Integrated Circuits Conference (CICC) Digest of Technical Papers*, Oct. 2004, pp. 95–98.
121. H. Kim and K. Roy, "Ultra-Low Power DLMS Adaptive Filter for Hearing Aid Applications," in *International Symposium on Low-Power Electronics and Design (ISLPED) Digest of Technical Papers*, Aug. 2001, pp. 352–357.
122. H. Soeleman, K. Roy, and B. Paul, "Sub-Domino Logic: Ultra-Low Power Dynamic Sub-Threshold Digital Logic," in *International Conference on VLSI Design (VLSI-Design) Digest of Technical Papers*, Jan. 2001, pp. 211–214.
123. K. Bowman, S. Duvall, and J. Meindl, "Impact of Die-to-Die and Within-Die Parameter Fluctuations on the Maximum Clock Frequency Distribution for Gigascale Integration," *IEEE Journal of Solid-State Circuits*, vol. 37, no. 2, pp. 183–190, February 2002.
124. A. Pelgrom, M. Duinmaijer and A. Welbers, "Matching Properties of MOS Transistors," *IEEE Journal of Solid-State Circuits*, vol. 24, no. 5, pp. 1433–1440, October 1989.
125. J. M. Rabaey, A. Chandrakasan, and B. Nikolic, *Digital Integrated Circuits: A Design Perspective*, 2nd ed.   Prentice Hall, 2003.
126. S. Agarwala, P. Wiley, A. Rajagopal, A. Hill, R. Damodaran, L. Nardini, T. Anderson, S. Mullinnix, J. Flores, Y. Heping, A. Chachad, J. Apostol, K. Castille, U. Narasimha, T. Wolf, N. Nagaraj, M. Krishnan, L. Nguyen, T. Kroeger, M. Gill, P. Groves, B. Webster, J. Graber, and C. Karlovich, "A 800 MHz system-on-chip for wireless infrastructure applications," in *Proceedings. 17th International Conference on VLSI Design*, 2004, pp. 381 – 389.
127. P. Royannez, H. Mair, F. Dahan, M. Wagner, M. Streeter, L. Bouetel, J. Blasquez, H. Clasen, G. Semino, J. Dong, D. Scott, B. Pitts, C. Raibaut, and U. Ko, "90nm low leakage SoC design techniques for wireless applications," in *IEEE International Solid-State Circuits Conference (ISSCC) Digest of Technical Papers*, 2005, pp. 138 – 589.
128. S. Hsu, B. Chatterjee, M. Sachdev, A. Alvandpour, R. Krishnamurthy, and S. Borkar, "A 90 nm 6.5 GHz 256/spl times/64 b dual supply register file

with split decoder scheme," in *Symposium on VLSI Circuits (VLSI) Digest of Technical Papers*, 2003, pp. 237 – 238.
129. M. Sumita and Y. Ikeda, "A 32b 64-word 9-read-port/7-write-port pseudo dual-bank register file using copied memory cells for a multi-threaded processor," in *IEEE International Solid-State Circuits Conference (ISSCC) Digest of Technical Papers*, Feb 2005, pp. 384 – 386.
130. K. Zhang, K. Hose, V. De, and B. Senyk, "The Scaling of Data Sensing Schemes for High Speed Cache Design in Sub-0.18μm," in *Symposium on VLSI Circuits (VLSI) Digest of Technical Papers*, 2000, pp. 226–227.
131. M. Yamaoka, Y. Shinozaki, N. Maeda, Y. Shimazaki, K. Kato, S. Shimada, K. Yanagisawa, and K. Osadal, "A 300MHz 25μA/Mb Leakage On-Chip SRAM Module Featuring Process-Variation Immunity and Low-Leakage-Active Mode for Mobile-Phone Application Processor," in *IEEE International Solid-State Circuits Conference (ISSCC) Digest of Technical Papers*, 2004, pp. 494–495.
132. N. Kim, K. Flautner, D. Blaauw, and T. Mudge, "Circuit and Microarchitectural Techniques for Reducing Cache Leakage Power," *IEEE Transactions on Very Large Scale Integration (VLSI) Systems*, vol. 12, no. 2, pp. 167–184, Feb. 2004.
133. A. Bhavnagarwala, S. Kosonocky, S. Kowalczyk, R. Joshi, Y. Chan, U. Srinivasan, and J. Wadhwa, "A Transregional CMOS SRAM with Single, Logic V_{DD} and Dynamic Power Rails," in *Symposium on VLSI Circuits (VLSI) Digest of Technical Papers*, 2004, pp. 292–293.
134. K. Itoh, "Low-Voltage Memories for Power-Aware Systems," in *International Symposium on Low-Power Electronics and Design (ISLPED) Digest of Technical Papers*, Aug., 2002, pp. 1–6.
135. E. Seevinck, F. List, and J. Lohstroh, "Static Noise Margin Analysis of MOS SRAM Cells," *IEEE Journal of Solid-State Circuits*, vol. SC-22, no. 5, pp. 748–754, Oct. 1987.
136. H. Qin, Y. Cao, D. Markovic, A. Vladimirescu, and J. Rabaey, "SRAM Leakage Suppression by Minimizing Standby Supply Voltage," in *International Symposium on Quality Electronic Design (ISQED) Digest of Technical Papers*, 2004, pp. 55–60.
137. K. Kanda, T. Miyazaki, M. K. Sik, H. Kawaguchi, and T. Sakurai, "Two Orders of Magnitude Leakage Power Reduction of Low Voltage SRAM's by Row-by-Row Dynamic V_{DD} Control (RRDV) Scheme," in *IEEE International ASIC/SOC Conference*, 2002, pp. 381–385.
138. T. Enomoto, Y. Oka, and H. Shikano, "A Self-Controllable Voltage Level (SVL) Circuit and its Low-Power High-Speed CMOS Circuit Applications," *IEEE Journal of Solid-State Circuits*, vol. 38, no. 7, pp. 1220–1226, 2003.
139. M. Yamaoka, K. Osada, and K. Ishibashi, "0.4-V Logic Library Friendly SRAM Array Using Rectangular-Diffusion Cell and Delta-Boosted-Array-Voltage Scheme," in *Symposium on VLSI Circuits (VLSI) Digest of Technical Papers*, 2002, pp. 170–173.
140. ——, "0.4-V Logic-Library-Friendly SRAM Array Using Rectangular-Diffusion Cell and Delta-Boosted-Array Voltage Scheme," *IEEE Journal of Solid-State Circuits*, vol. 39, no. 6, pp. 934–940, 2004.
141. A. Bhavnagarwala, A. Kapoor, and J. Meindl, "Dynamic-Threshold CMOS SRAM Cells for Fast, Portable Applications," in *IEEE International ASIC/SOC Conference*, 2000, pp. 359–363.

202   References

142. K. Itoh, A. Fridi, A. Bellaouar, and M. Elmasry, "A Deep Sub-V, Single Power-Supply SRAM Cell with Multi-V_T, Boosted Storage Node and Dynamic Load," in *Symposium on VLSI Circuits (VLSI) Digest of Technical Papers*, 1996, pp. 132–133.
143. H. Yamauchi, T. Iwata, H. Akamatsu, and A. Matsuzawa, "A 0.8V/100MHz/sub-5mW-operated Mega-Bit SRAM Cell Architecture with Charge-Recycle Offset-Source Driving (OSD) Scheme," in *Symposium on VLSI Circuits (VLSI) Digest of Technical Papers*, 1996, pp. 126–127.
144. K. Osada, Y. Saitoh, E. Ibe, and K. Ishibashi, "16.7-fA/Cell Tunnel-Leakage-Suppressed 16-Mb SRAM for Handling Cosmic-Ray-Induced Multierrors," *IEEE Journal of Solid-State Circuits*, vol. 38, no. 11, pp. 1952–1957, Nov. 2003.
145. K. Zhang, U. Bhattacharya, Z. Chen, F. Hamzaoglu, D. Murray, N. Vallepalli, Y. Yang, B. Zheng, and M. Bohr, "A SRAM Design on 65nm CMOS Technology with Integrated Leakage Scheme," in *Symposium on VLSI Circuits (VLSI) Digest of Technical Papers*, 2004, pp. 294–295.
146. K. Nii, Y. Tsukamoto, T. Yoshizawa, S. Imaoka, and H. Makino, "A 90nm Dual-Port SRAM with 2.04μm^2 8T-Thin Cell Using Dynamically-Controlled Column Bias Scheme," in *IEEE International Solid-State Circuits Conference (ISSCC) Digest of Technical Papers*, 2004, pp. 508–509.
147. A. Agarwal, H. Li, and K. Roy, "A Single-V_t Low-Leakage Gated-Ground Cache for Deep Submicron," *IEEE Journal of Solid-State Circuits*, vol. 38, no. 2, pp. 319–328, Feb. 2003.
148. K. Kanda, H. Sadaaki, and T. Sakurai, "90% Write Power-Saving SRAM Using Sense-Amplifying Memory Cell," *IEEE Journal of Solid-State Circuits*, vol. 39, no. 6, pp. 927–933, 2004.
149. K. Osada, J. L. Shin, M. Khan, Y. Liou, K. Wang, K. Shoji, K. Kuroda, S. Ikeda, and K. Ishibashi, "Universal-Vdd 0.65–2.0-V 32-kB Cache Using a Voltage-Adapted Timing-Generation Scheme and a Lithographically Symmetrical Cell," *IEEE Journal of Solid-State Circuits*, vol. 36, pp. 1738–1744, Nov. 2001.
150. S. Ikeda, Y. Yoshida, K. Ishibashi, and Y. Mitsui, "Failure Analysis of 6T SRAM on Low-Voltage and High-Frequency Operation," *IEEE Transactions on Electron Devices*, vol. 50, no. 5, pp. 1270–1276, May 2003.
151. C. Lage, D. Burnett, T. McNelly, K. Baker, A. Bormann, D. Dreier, and V. Soorholtz, "Soft Error Rate and Stored Charge Requirements in Advanced High-Density SRAMs," in *International Electron Devices Meeting (IEDM) Digest of Technical Papers*, 1993, pp. 821–824.
152. K. Osada, K. Yamaguchi, Y. Saitoh, and T. Kawahara, "SRAM Immunity to Cosmic-Ray-Induced Multierrors Based on Analysis of an Induced Parasitic Bipolar Effect," *IEEE Journal of Solid-State Circuits*, vol. 39, no. 5, pp. 827–833, May 2004.
153. H. Kawaguchi, Y. Itaka, and T. Sakurai, "Dynamic Cut-off Scheme for Low-Voltage SRAM's," in *Symposium on VLSI Circuits (VLSI) Digest of Technical Papers*, 1998, pp. 140–141.
154. M. Yamaoka, N. Maeda, Y. Shinozaki, Y. Shimazaki, K. Nii, S. Shimada, K. Yanagisawa, and T. Kawahara, "Low-Power Embedded SRAM Modules with Expanded Margins for Writing," in *IEEE International Solid-State Circuits Conference (ISSCC) Digest of Technical Papers*, Feb. 2005, pp. 480–481.

155. M. Yamaoka, K. Osada, R. Tsuchiya, M. Horiuchi, S. Kimura, and T. Kawahara, "Low Power SRAM Menu for SOC Application Using Yin-Yang-Feedback Memory Cell Technology," in *Symposium on VLSI Circuits (VLSI) Digest of Technical Papers*, 2004, pp. 288–291.
156. B. Calhoun and A. Chandrakasan, "Analyzing Static Noise Margin for Subthreshold SRAM in 65nm CMOS," in *Proceedings of the European Solid-State Circuits Conference*, 2005, pp. 363–366.
157. K. Takeda, Y. Hagihara, Y. Aimoto, M. Nomura, Y. Nakazawa, T. Ishii, and H. Kobatake, "A Read-Static-Noise-Margin-Free SRAM Cell for Low-Vdd and High-Speed Applications," in *IEEE International Solid-State Circuits Conference (ISSCC) Digest of Technical Papers*, Feb. 2005, pp. 478–479.
158. A. Alvandpour, D. Somasekhar, R. Krishnamurthy, V. De, S. Borkar, and C. Svensson, "Bitline Leakage Equalization for Sub-100nm Caches," in *Proceedings of the European Solid-State Circuits Conference*, 2003, pp. 401–404.
159. E. Vittoz, "MOS Transistors Operated in the Lateral Bipolar Mode and Their Applications in CMOS Technology," *IEEE Journal of Solid-State Circuits*, vol. 18, no. 6, pp. 273–279, June 1983.
160. B. Minch, "A Low-Voltage MOS Cascode Bias Circuit for All Current Levels," in *International Symposium on Circuits and Systems (ISCAS) Digest of Technical Papers*, vol. 3, Sydney, 2002, pp. 619–622.
161. T. Choi, R. Kaneshiro, R. Brodersen, P. Gray, W. Jett, and M. Wilcox, "High-frequency CMOS Switched Capacitor Filters for Communications Applications," *IEEE Journal of Solid-State Circuits*, vol. 18, pp. 652–664, Dec. 1983.
162. A. Porret, T. Melly, D. Python, C. C. Enz, and E. A. Vittoz, "An Utralow-Power UHF Transceiver Integrated in a Standard Digital CMOS Process: Architecture and Receiver," *IEEE Journal of Solid-State Circuits*, vol. 36, pp. 462–466, March 2001.
163. V. von Kaenel, M. D. Pardoen, E. Dijkstra, and E. Vittoz, "Automatic Adjustment of Threshold and Supply Voltages for Mnimum Power Consumption in CMOS Digital Circuits," in *International Symposium on Low-Power Electronics and Design (ISLPED) Digest of Technical Papers*, San Diego, October 1994, pp. 78–79.
164. F. Krummenacher and N. Joehl, "A 4-MHz CMOS Continuous-Time Filter with On-Chip Automatic Tuning," *IEEE Journal of Solid-State Circuits*, vol. 23, pp. 750–758, June 1988.
165. B. Gilbert, "The Multi-Tanh Principle: A Tutorial Review," *IEEE Journal of Solid-State Circuits*, vol. 33, pp. 2–7, January 1998.
166. H. J. Oguey and D. Aebischer, "CMOS Current Reference Without Resistance," *IEEE Journal of Solid-State Circuits*, vol. 32, pp. 1132–1135, July 1997.
167. M. Dutoit and F. Sollberger, "Lateral Polysilicon pn Diodes," *Journal of Electrochemical Society*, vol. 125, pp. 1648–1651, October 1978.
168. V. von Kaenel and E. Vittoz, "Crystal Oscillators," in *Analog Circuit Design*. Kluwer, 1996.
169. D. Aebischer, H. J. Oguey, and V. von Kaenel, "A 2.1 MHz Crystal Oscillator Time Base With a Current Consumption Under 500nA," *IEEE Journal of Solid-State Circuits*, vol. 32, pp. 999–1005, July 1997.
170. W. Thommen, "An Improved Low Power Crystal Oscillator," in *Proceedings of the European Solid-State Circuits Conference*, Sept. 1999, pp. 146–149.
171. B. Gilbert, "Translinear Circuits: a Proposed Classification," *El. Letters*, vol. 11, p. 14, 1975.

172. E. Vittoz, "Analog VLSI Implementation of Neural Networks," in *Handbook of Neural Computation*. Oxford Univerity Press, 1996.
173. B. Gilbert, "A New Wide-Band Amplifier Technique," *IEEE Journal of Solid-State Circuits*, vol. 3, pp. 353–365, December 1968.
174. ——, *Translinear Circuits*, private ed. Barrie Gilbert, 1981.
175. A. Andreou and K. Boahen, "Neural Information Processing II," in *Analog VLSI Signal and Information Processing*, M. Ismail and T. Fiez, Eds. McGraw-Hill, 1994.
176. E. Seevinck, E. Vittoz, T. H. Joubert, and W. Beetge, "CMOS Translinear Circuits for Minimum Supply Voltage," *IEEE Trans. Circuits and Systems II*, vol. 47, pp. 1560–1564, December 2000.
177. R. W. Adams, "Filtering in the Log Domain," presented at the 63rd AES Audio Engineering Soc. Conf., NewYork, May 1979.
178. E. Seevinck, "Companding Current-Mode Integrator: A New Circuit Principle for Continuous-Time Monolithic Filters," *El. Letters*, vol. 26, no. 24, pp. 2046–2047, November 1990.
179. D. R. Frey, "Log-Domain Filtering: An Approach to Current Mode Filtering," *Proc. IEE*, vol. 140, pp. 406–416, December 1993.
180. C. Enz, M. Punzenberger, and D. Python, "Low-Voltage Log-Domain Signal Processing in CMOS and BiCMOS," *IEEE Trans. Circuits Syst. II*, vol. 46, pp. 279–289, March 1999.
181. M. Punzenberger and C. Enz, "A 1.2-V Low-Power BiCMOS Class-AB Log-Domain Filters," *IEEE Journal of Solid-State Circuits*, vol. 32, pp. 1968–1978, December 1997.
182. D. Python and C. Enz, "A Micropower Class-AB CMOS Log-Domain Filter for DECT Applications," *IEEE Journal of Solid-State Circuits*, vol. 36, no. 7, pp. 1067–1075, July 2001.
183. N. Krishnapura and Y. Tsividis, "Micropower Low-Voltage Analog Filter in a Digital CMOS Process," *IEEE Journal of Solid-State Circuits*, vol. 38, no. 6, pp. 1063–1067, June 2003.
184. T. Delbruck, "Bump Circuits for Computing Similarity and Dissimilarity of Analog Voltages," in *Proc. Int. Joint Conf. on Neural Networks*, vol. 1, 1991, pp. I-475,I-479.
185. P. Venier, "A Contrast Sensitive Silicon Retina Based on Conductance Modulation in a Diffusion Network," in *Proc. of MicroNeuro 97*, Dresden, 1997.
186. E. Vittoz, "Present and Future Industrial Applications of Bio-Inspired VLSI Systems," *Analog Integrated Circuits and Signal Processing*, vol. 30, pp. 173–184, 2002.
187. Z. Toprak Deniz, "Multi-Unit Global Energy Management and Optimization for Network-on-Chip Applications," Ph.D. dissertation, EPFL, Lausanne, Switzerland, 2006.
188. J. C. Maxwell, *A Treatise in Electricity and Magnetism*, 3rd ed. Dover Publications, Inc., 1954.
189. A. Wang and A. Chandrakasan, "A 180-mV subthreshold FFT processor using a minimum energy design methodology," *IEEE Journal of Solid-State Circuits*, vol. 40, no. 1, pp. 310–319, Jan 2005.
190. B. M. Baas, "A low-power, high-performance, 1024-point FFT processor," *IEEE Journal of Solid-State Circuits*, vol. 34, no. 3, pp. 380–387, Mar. 1999.

191. P. Macken, M. Degrauwe, M. V. Paemel, and H. Oguey, "A Voltage Reduction Technique for Digital Systems," in *IEEE International Solid-State Circuits Conference (ISSCC) Digest of Technical Papers*, Feb. 1990, pp. 238–239.
192. [Online]. Available: http://www.intel.com/design/intelxscale/
193. K. Nowka, G. Carpenter, E. MacDonald, H. Ngo, B. Brock, K. Ishii, T. Nguyen, and J. Burns, "A 0.9V to 1.95V Dynamic Voltage-Scalable and Frequency-Scalable 32b PowerPC Processor," in *IEEE International Solid-State Circuits Conference (ISSCC) Digest of Technical Papers*, Feb. 2002, pp. 340–341.
194. [Online]. Available: http://www.transmeta.com/crusoe/index.html
195. H. Kawaguchi, G. Zhang, S. Lee, and T. Sakurai, "A Controller LSI for Realizing VDD-Hopping Scheme with Off-the-Shelf Processors and Its Application to MPEG4 System," *IEICE Transactions on Electronics*, vol. E85-C, no. 2, pp. 263–271, Feb. 2002.
196. B. Calhoun and A. Chandrakasan, "Ultra-Dynamic Voltage Scaling (UDVS) Using Sub-threshold Operation and Local Voltage Dithering in 90nm CMOS," in *IEEE International Solid-State Circuits Conference (ISSCC) Digest of Technical Papers*, Feb. 2005, pp. 300–301.
197. A. Chandrakasan, V. Gutnik, and T. Xanthopoulos, "Data Driven Signal Processing: An Approach for Energy Efficient Computing," in *International Symposium on Low-Power Electronics and Design (ISLPED) Digest of Technical Papers*, 1996, pp. 347–352.
198. L. Chandrasena, P. Chandrasena, and M. Liebelt, "An Energy Efficient Rate Selection Algorithm for Voltage Quantized Dynamic Voltage Scaling," in *International Symposium on System Synthesis*, Oct. 2001, pp. 124–129.
199. S. Lee and T. Sakurai, "Run-time Power Control Scheme Using Software Feedback Loop for Low-Power Real-time Applications," in *Asia and South-Pacific Design Automation Conference*, Jan. 2000, pp. 381–386.
200. ——, "Run-time Voltage Hopping for Low-power Real-time Systems," in *ACM/IEEE Design Automation Conference (DAC) Digest of Technical Papers*, 2000, pp. 806–809.
201. H. Kawaguchi, K. Kanda, K. Nose, S. Hattori, D. Dwi, D. Antono, D. Yamada, T. Miyazaki, K. Inagaki, T. Hiramoto, and T. Sakurai, "A 0.5V, 400MHz, VDD-Hopping Processor with Zero-VTH FD-SOI Technology," in *IEEE International Solid-State Circuits Conference (ISSCC) Digest of Technical Papers*, Feb. 2003, pp. 106–107.
202. L. Chandrasena and M. Liebelt, "Energy Minimization in Dynamic Supply Voltage Scaling Systems Using Data Dependent Voltage Level Selection," in *International Symposium on Circuits and Systems (ISCAS) Digest of Technical Papers*, 2000, pp. 525–528.

Index

above-threshold, 32, 35, 40, 77, 88, 181
accumulator, 179
activity factor, 21, 28, 172
adder, 15, 22, 96
   Kogge-Stone, 16, 179
ambient energy, 5
amplitude regulator, 157
analog circuits, 147–166

battery, 1
   lifetime, 2–4, 168
bitcell, 104
body-bias, 20–22, 152

cascode mirror, 148
CMOS circuits, 7, 11, 14, 15, 19, 75–91, 93, 96
   sub-threshold, 22, 75, 103–114, 167
current
   conduction, 54
   diffusion, 7, 30, 54
   drain, 54
   forward, 55
   reverse, 55
   short-circuit, 40
current mirror, 148, 156
current reference, 156
current-mode multiplier/divider, 159

deep sub-micron, 13, 92
device
   mismatch, 94
DIBL, 11, 30, 33, 34, 39, 72, 78, 151
differential pair, 152, 154

DLMS filter, 22
drain-to-source current, 77, 79
DSP, 3, 15, 16, 103, 168
DVS, 16, 174
dynamic circuits, 92, 106
dynamic energy, 27, 36
dynamic logic, 99

EKV, 9, 49–74
electric field, 11
energy, 25, 36
   harvesting, 4
energy-aware architectures, 169
energy-constrained, 2, 36
energy-delay product, 15, 21
energy-performance contours, 25–28, 30
   FFT, 170
   ring oscillator, 26, 28

fast Fourier Transform(FFT), 39, 114, 167–172
   complex-valued, 168
   real-valued, 168
FFT, 39
FIR filter, 16, 39, 42, 83, 89, 90, 174
flip-flop, 87, 88

gate leakage, 34, 35, 71
gate oxide, 34
GIDL, 34, 111

header switch, 179, 183
hearing-aid, 22

Intel

208 Index

Pentium, 16
StrongArm, 16
inversion coefficient, 57, 68, 147, 149
inverter, 18, 32, 35, 75
  delay, 75
  load lines, 77
  voltage transfer curves, 77
ITRS, 11, 14

Lambert W, 37, 46
latch, 89
leakage energy, 18, 27, 36
LFSR, 90
local voltage dithering(LVD), 177
logic families, 92–102
low-pass filter, 158

MCU, 3
minimum energy, 1, 19, 20, 27, 30, 35, 38, 180
minimum voltage, 17, 79, 84, 90
MIPS, 3
mismatch, 159
  threshold, 156
model, 30, 32–74
  noise, 63
  small-signal, 59
  small-signal AC, 61
  transconductance, 59
  transregional, 21
monte-carlo, 94
MPEG video processing, 174
MTCMOS, 179
multiplier, 166
  Baugh-Wooley, 169
  variable bit-precision, 169
multiply-accumulate, 20
multistage variable linear attenuator, 164
mux, 113

nand, 86
noise, 63, 151
  channel, 64
  interface, 65

operational transconductance amplifier, 154
optimum energy, 26, 35, 36, 38

optimum supply voltage, 25
optimum threshold voltage, 25, 35
optimum voltage supply, 35
oscillator, 18, 22, 25, 39, 87, 90, 157, 166, 179

parallelism, 15, 46, 84, 89
pipelining, 15, 46
power-delay product, 12, 17, 19, 92
Process Variations, 88
process variations, 19, 80, 82–103
pseudo-nMOS, 22, 92, 109
pseudo-resistor, 162, 163

quasi-linear, 32

ratioed circuits, 88, 95
  flip-flop, 88
RBB, 34, 35
read bitline, 105
  hierarchical, 113
  negative wordline, 110
  precharge, 106
  pseudo-nMOS, 109
  tristate read, 111
read-only memory(ROM), 169, 172
register file, 103–114, 169, 172
RFID, 3

saturation voltage, 58
sense-amplifier, 105
short-channel effects, 63, 74
sizing, 79, 82, 88, 91, 95
  header switch, 179
  minimum, 91
  minimum energy, 82
SNM, 115–126, 129–131, 133, 139, 144–146
SRAM, 15, 115–146
  read, 133
  write, 137
stacking, 86
standard cells, 83–91
static CMOS, 84
strong inversion, 18, 22, 32, 53, 56, 69, 84, 104, 181
sub-threshold model, 8, 30, 32–74
sub-threshold slope, 8, 76
switching energy, 18, 27, 36

temperature, 35, 45, 46, 65, 89, 157, 159, 180
Texas Instruments
  C5xx, 3, 16
  C64x, 103
  MSP430, 3
  OMAP2, 103
thermal voltage, 17
threshold function, 50
threshold voltage, 52
  variation, 19, 65, 69, 92, 94
transistor
  band gap widening, 70
  depletion capacitance, 59
  forward saturation, 57
  gate depletion, 69
  linear, 57
  mismatch, 67, 148, 151
  mobility, 45, 54, 157, 180
  noise, 63
  pseudo-resistor, 62, 152
  reliability, 12
  reverse saturation, 58
  saturation, 61, 150
  transconductance, 150
  weak inversion, 58
translinear circuits, 158
transmission gate, 96
tristate, 111
  C2MOS, 104
  inverter, 104

UDVS, 16, 173–185, 187, 188, 190
  test chip, 179

variable bit-precision, 169
variable activity factor, 25, 28
variable resistive networks, 163
velocity saturation, 32
voltage amplifier, 148
  low-voltage, 150
voltage dithering, 176
voltage follower, 155
voltage reference, 156
VTC, 77
  sub-threshold, 18, 82

weak inversion, 1, 7, 18, 22, 49, 54, 56, 58, 59, 69
wireless sensors, 2, 167, 169
workload, 42–44
wristwatch, 9, 22
write port, 104

XOR, 84

yield, 99, 101

DISCARDED
CONCORDIA UNIV. LIBRARY